THE **GORILLA GUIDE** TO...

Hypercor Infrastru Implementation Strategies

GW01465756

Scott D. Lowe • David M. Davis

ActualTech Media

The Gorilla Guide to Hyperconverged Infrastructure Implementation Strategies

Author:	Scott D. Lowe, ActualTech Media
Editors:	David M. Davis, ActualTech Media
	Hilary Kerchner, Dream Write Creative
Book Design:	Braeden Black, Avalon Media Productions
	Geordie Carswell, ActualTech Media
Layout:	Braeden Black, Avalon Media Productions

Printed in the United States of America

First Printing, 2015

ISBN 978-1-943952-00-7

ActualTech Media
Okatie Village Ste 103-157
Bluffton, SC 29909
www.actualtechmedia.com

Table of Contents

Section 1: Technical Overview

Section 2: Use Cases

Section 3: Organizational Considerations

About the Author

Scott D. Lowe, vExpert

Scott Lowe is a vExpert and partner and Co-Founder of ActualTech Media. Scott has been in the IT field for close to twenty years and spent ten of those years in filling the CIO role for various organizations. Scott has written thousands of articles and blog postings and regularly contributes to *www.EnterpriseStorageGuide.com* & www.*ActualTech.io.*

About the Editor

David M. Davis, vExpert

David Davis is a partner and co-founder of ActualTech Media. With over 20 years in enterprise technology, he has served as an IT Manager and has authored hundreds of papers, ebooks, and video training courses. He's a 6 x vExpert, VCP, VCAP, & CCIE# 9369. You'll find his vSphere video training at *www.Pluralsight.com* and he blogs at *www.VirtualizationSoftware.com* and www.*ActualTech.io.*

About ActualTech Media

ActualTech Media provides enterprise IT decision makers with the information they need to make informed, strategic decisions as they modernize and optimize their IT operations.

Leading 3rd party IT industry influencers Scott D. Lowe, David M. Davis and special technical partners cover hot topics from the software-defined data center to hyperconvergence and virtualization.

Cutting through the hype, noise and claims around new data center technologies isn't easy, but ActualTech Media helps find the signal in the noise. Analysis, authorship and events produced by ActualTech Media provide an essential piece of the technology evaluation puzzle.

More information available at **www.actualtechmedia.com**

ActualTech Media

simplivity™

About SimpliVity

SimpliVity hyperconverged infrastructure delivers the enterprise-class performance and availability today's IT leaders require, with the cloud economics businesses demand. No other company has taken on the mega task of assimilating all IT elements below the hypervisor into x86 building blocks that combine to create a single, scalable resource pool. SimpliVity's unique data architecture improves performance and data efficiency, and enables built-in data protection and global unified management from a single management interface. SimpliVity's hyperconverged infrastructure simplifies IT infrastructure, reducing CapEx; accelerates deployment and streamlines operations, saving run-rate expenses; and improves recovery objectives, minimizing risk—all contributing to a three-fold TCO savings.

Visit *www.SimpliVity.com*

Gorilla Guide Features

In the Book

These help point readers to other places in the book where a concept is explored in more depth.

The How-To Corner

These will help you master the specific tasks that it takes to be proficient in the technology jungle.

Food For Thought

In the these sections, readers are served tasty morsels of important information to help you expand your thinking .

School House

This is a special place where readers can learn a bit more about ancillary topics presented in the book.

Bright Idea

When we have a great thought, we express them through a series of grunts in the Bright Idea section.

Dive Deep

Takes readers into the deep, dark depths of a particular topic.

Executive Corner

Discusses items of strategic interest to business leaders.

SECTION 1

Technical Overview

1

Introduction to Hyperconverged Infrastructure

In recent years, it seems like technology is changing faster than it used to in decades past. As employees devour newer technologies such as smartphones, tablets, wearables, and other devices, and as they become more comfortable with solutions such as Dropbox and Skype, their demands on enterprise IT intensify. Plus, management and other decision makers are also increasing their demands on enterprise IT to provide more infrastructure with less cost and time. Unfortunately, enterprise IT organizations often don't see much, if any, associated increases in funding to accomodate these demands.

These demands have resulted in the need for IT organizations to attempt to mimic NASA's much-heralded "Faster, Better, Cheaper" operational campaign. As the name suggests, NASA made great attempts to build new missions far more quickly than was possible in the past, with greater levels of success, and with costs that were dramatically lower than previous missions. NASA was largely successful in their efforts, but the new missions tended to look very different from the ones in the past. For example, the early missions were big and complicated with a ton of moving parts, while modern missions have been much smaller in scale with far more focused mission deliverables.

What is NASA?

NASA is the United States National Aeronautical and Space Administration and has been responsible for helping the U.S. achieve success in its space programs, from the moon landing to recent high quality photographs of Pluto. NASA has faced serious budget cuts in recent years, but has been able to retool itself around smaller, more focused missions that cost less and have achieved incredible results.

The same "faster, better, cheaper" challenge is hitting enterprise IT, although even the hardest working IT pros don't usually have to make robots rove the surface of an inhospitable planet! Today's IT departments must meet a growing list of business needs while, at the same time, appeasing the decision makers who demand far more positive economic outcomes (either by cutting costs overall or doing more work within the existing budget).

Unfortunately, most of today's data center architectures actively work against these goals, because with increasing complexity comes increased costs — and things have definitely become more complex. Virtualization has been a fantastic opportunity for companies, but with virtualization has come some new challenges, including major issues with storage. With virtualization, enterprise IT has moved from physical servers, where storage services could be configured on a per-server basis, to shared storage systems. These shared storage systems, while offering plenty of capacity, have often not been able to keep up in terms of performance, forcing IT departments to take corrective actions that don't always align with good economic practices. For example, it's common for IT pros to add entire shelves of disks, not because they need the capacity, but because they need the spindles to increase overall storage performance. There are, of course, other ways to combat storage performance issues, such as through the use of solid state disk (SSD) caching systems, but these also add complexity to what is already a complex situation.

There are other challenges that administrators of legacy data centers need to consider as well:

- **Hardware sprawl.** Data centers are littered with separate infrastructure silos that are all painstakingly cobbled together to form a complete solution. This hardware sprawl results in a data center that is increasingly complex, decreasing flexibility, and expensive to maintain.

- **Policy sprawl.** The more variety of solutions in the data center, the more touch points that exist when it comes to applying consistent policies across all workloads.

- **Scaling challenges.** Predictability is becoming really important. That is, being able to predict ongoing budgetary costs and how well a solution will perform after purchase are important. Legacy infrastructure and its lack of inherent feature-like scaling capability make both predictability metrics very difficult to achieve.

- **Desire for less technical overhead.** Businesses want analysts and employees that can help drive top line revenue growth. Purely technical staff are often considered expenses that must be minimized. Businesses today are looking for ways to make the IT function easier to manage overall so that they can redeploy technical personnel to more business-facing needs. Legacy data centers are a major hurdle in this transition.

So, with all of this in mind, what are you to do?

Hyperconverged Infrastructure from 30,000 Feet

An emerging data center architectural option, dubbed *hyperconverged infrastructure*, is a new way to reduce your costs and better align enterprise IT with business needs. At its most basic, hyperconverged infrastructure is the conglomeration of the servers and storage devices that comprise the data center. These systems are wrapped in comprehensive and easy-to-use management tools designed to help shield the administrator from much of the underlying architectural complexity.

Why are these two resources, storage and compute, at the core of hyperconverged infrastructure? Simply put, storage has become an incredible challenge for many companies. It's one of— if not *the* — most expensive resources in the data center and often requires a highly skilled person or team to keep it running. Moreover, for many companies, it's a single point of failure. When storage fails, swaths of services are negatively impacted.

Combining storage with compute is in many ways a return to the past, but this time many new technologies have been wrapped around it. Before virtualization and before SANs, many companies ran physical servers with directly attached storage systems, and they tailored these storage systems to meet the unique needs for whatever applications might have been running on the physical servers. The problem with this approach was it created numerous "islands" of storage and compute resources. Virtualization solved this resource-sharing problem but introduced its own problems previously described.

Hyperconverged infrastructure distributes the storage resource among the various nodes that comprise a cluster. Often built using commodity server chasses and hardware, hyperconverged infrastructure nodes and appliances are bound together via Ethernet and a powerful software layer. The software layer often includes a *virtual storage appliance* (VSA) that runs on each cluster node. Each VSA then communicates with all of the other VSAs in the cluster over an Ethernet link, thus forming a distributed file system across which virtual machines are run.

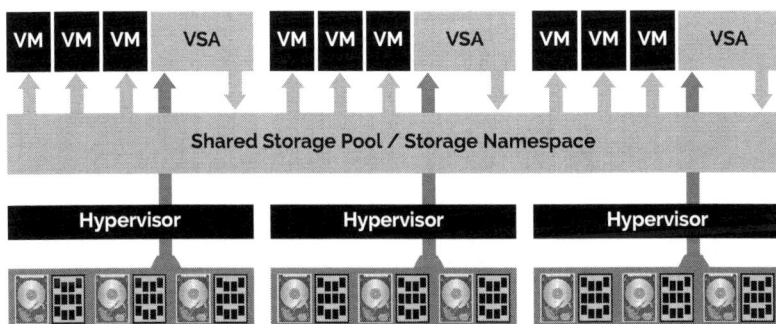

Figure 1-1: An overview of a Virtual Storage Appliance

The fact that these systems leverage commodity hardware is critical. The power behind hyperconverged infrastructure lies in its ability to coral resources – RAM, compute, and data storage – from hardware that doesn't all have to be custom-engineered. This is the basis for hyperconverged infrastructure's ability to scale granularly and the beginnings of cost reduction processes.

The basics behind hyperconverged infrastructure should be well understood before proceeding with the remainder of this book. If you're new to hyperconverged infrastructure or are unfamiliar with the basics, please read *Hyperconverged Infrastructure for Dummies*, available now for free from www.hyperconverged.org.

Resources to Consolidate

The basic combination of storage and servers is a good start, but once one looks beyond the confines of this baseline definition, hyperconverged infrastructure begins to reveal its true power. The more hardware devices and software systems that can be collapsed into a hyperconverged solution, the easier it becomes to manage the solution and the less expensive it becomes to operate.

Here are some data center elements that can be integrated in a hyperconverged infrastructure.

Deduplication Appliances

In order to achieve the most storage capacity, deduplication technologies are common in today's data center. Dedicated appliances are now available which handle complex and CPU-intensive deduplication tasks, ultimately reducing the amount of data that has to be housed on primary storage. Deduplication services are also included with storage arrays in many cases. However, deduplication in both cases is not as comprehensive as it could be. As data moves around the organization, data is rehydrated into its original form and may or may not be reduced via deduplication as it moves between services.

SSD Caches/All-Flash Array

To address storage performance issues, companies sometimes deploy either solid state disk (SSD)-based caching systems or full SSD/flash-based storage arrays. However, both solutions have the potential to increase complexity as well as cost. When server-side PCI-e SSD cards are deployed, there also has to be a third-party software layer that allows them to act as a cache, if that is the desire. With all-flash arrays or flash-based stand-alone caching systems, administrators are asked to support new hardware in addition to everything else in the data center.

Backup Software

Data protection in the form of backup and recovery remains a critical task for IT and is one that's often not meeting organizational needs. Recovery time objectives (RTO) and recovery point objectives (RPO) — both described in the deep dive section entitled "The Ins and Outs of Backup and Recovery" — are both shrinking metrics that IT needs to improve upon. Using traditional hardware and software solutions to meet this need has been increasingly challenging. As RPO and RTO needs get shorter, costs get higher with traditional solutions.

With the right hyperconverged infrastructure solution, the picture changes a bit. In fact, included in some baseline solutions is a comprehensive backup and recovery capability that can enable extremely short RTO windows while also featuring very small RPO metrics.

The Ins & Outs of Backup & Recovery

There are critical recovery metrics – known as Recovery Time Objective (RTO) and Recovery Point Objective (RTO) that must be considered in your data protection plans. You can learn a lot more about these two metrics in Chapter 4.

Data Replication

Data protection is about far more than just backup and recovery. What happens if the primary data center is lost? This is where replicated data comes into play. By making copies of data and replicating that data to remote sites, companies can rest assured that critical data won't be lost.

To enable these data replication services, companies implement a variety of other data center services. For example, to minimize replication impact on bandwidth, companies deploy WAN acceleration devices intended to reduce the volume of data traversing the Internet to a secondary site. WAN accelerators are yet another device that needs to be managed, monitored, and maintained. There are acquisition costs to procure these devices; there are costs to operate these devices in the form of staff time and training; and there are annual maintenance costs to make sure that these devices remain supported by the vendor.

Up Next

With an understanding of hyperconverged infrastructure and knowledge about many of the resources that can be consolidated into such solutions, let's move on to discuss some specific data center architectural elements and options that comprise a hyperconverged environment.

2

Architecting the Hyperconverged Data Center

Data centers are dynamic, complex, and sometimes even chaotic. As business needs evolve, so does the data center, with IT staff working hard to ensure that the operating environment is sufficiently robust. Hyperconverged infrastructure starts to change the mechanics behind how these efforts are carried out. With regard to hyperconvergence, there are a number of architectural elements that must be considered in order to determine the best path forward. But always remember: one of the primary goals of hyperconvergence is to simplify infrastructure decisions in the data center.

You don't need to worry about buying all kinds of different hardware, because with hyperconvergence the traditional silos of compute and storage resources can be wrapped up into a single hyperconverged appliance. Moreover, with the right hyperconverged infrastructure solution, you can converge far more than just servers and storage. You can also include your entire backup-and-recovery process, your deduplication and WAN acceleration appliances, and much more. Your architectural decisions can revolve around higher-order items, such as those described in the following sections.

Decision 1: Server Support

Not all hyperconverged solutions ship in the same kind of packaging. For example, there are appliance-based hyperconverged solutions from companies such as SimpliVity, Nutanix, Scale Computing, and Maxta. And then there are software-only solutions that you install yourself, which include Stratoscale and Maxta. Maxta is on both lists because they support both pre-configured appliances and software-only.

With an appliance-based solution, you're buying the full package, and you just need to plug everything in and turn it on. These are really easy to get going since most things are already done for you. However, with an appliance-based solution, you generally have to live with whatever constraints the vendor has placed on you. You need to remain within their hardware specifications, and you don't always get to choose your server platform, although many appliance-based solutions do support servers from multiple vendors. For example, SimpliVity solutions can be shipped on SimpliVity's Dell server platform or on Cisco UCS thanks to SimpliVity's partnership with Cisco. Nutanix can be purchased on either Dell or Supermicro, and Maxta has relationships with a variety of server vendors.

If you'd rather go your own way with regard to hardware, you can choose a software-based hyperconverged solution. With these products, you buy your own server hardware and configure what you want for each resource, making sure to stay within the minimum guidelines required by the hyperconverged infrastructure solution. Once you have the server hardware delivered, you install the hyperconverged infrastructure software and configure it to meet your needs.

Software-based solutions are really good for larger organizations with sufficient staff to install and support the hyperconverged infrastructure. Hardware-based solutions are often desired by companies that are looking for a more seamless deployment experience or that do not have sufficient staff to handle these tasks.

Decision 2: The Storage Layer

Let's face facts. One of the main reasons people are dissatisfied with their data centers is because their storage solution has failed to keep pace with the needs of the business. It's either too slow to support mission-critical applications or it doesn't have data efficiency features (deduplication and compression), thus forcing the company to buy terabyte after terabyte of new capacity.

Many storage devices are not well-designed when it comes to supporting virtualized workloads, either. Traditional SANs are challenged when attempting to support the wide array of I/O types that are inherent in heavily-virtualized environments. At the same time, storage has become more complex, often requiring specialized skill sets to keep things running. For some systems, it's not easy to do the basics, which can include managing LUNs, RAID groups, aggregates and more.

As companies grow and become more dependent on IT, they also start to have more reliance on data mobility services. Legacy storage systems don't always do a great job enabling data mobility and often don't even support services like remote replication and cloning or, if they do, it's a paid upgrade service. Without good local and remote cloning and replication capabilities, ancillary needs like data protection take on new challenges, too.

None of these situations are sustainable for the long term, but companies have spent inordinate sums of cash dragging inadequate storage devices into the future.

Hyperconverged infrastructure aims to solve this storage challenge once and for all. At the most basic level, hyperconverged infrastructure unifies the compute and storage layers and effectively eliminates the need for a monolithic storage array and SAN.

How does the storage component actually work if there is no more SAN? Let's unveil the storage secrets you've been dying to know.

Software-Defined Storage Defined

Abstract. Pool. Automate. That is the mantra by which the software-defined movement attains its success. Consider the SAN. It's a huge and expensive device. Software-defined storage (SDS) works in a vastly different way. With SDS, storage resources are abstracted from the underlying hardware. In essence, physical storage resources are logically separated from the system via a software layer.

Hyperconverged infrastructure systems operate by returning to an IT environment that leverages direct-attached storage running on commodity hardware, but many solutions go far beyond this baseline. In these baseline systems, there are a multitude of hard drives and solid state disks installed in each of the x86-based server nodes that comprise the environment. Installed on each of these nodes is the traditional hypervisor along with software to create a shared resource pool of compute and storage.

What's more is that there are vendors who collapse data protection, cloud gateway technologies, and services such as deduplication, compression and WAN optimization into their solution. In essence, hyperconverged infrastructure leverages the concepts behind software-defined storage systems in order to modernize and simplify the data center environment.

With storage hardware fully abstracted into software, it becomes possible to bring policy-based management and APIs to bear in ways that focus efforts on management on the virtual machine rather than the LUN. The virtual machine (VM) is really the administrative target of interest whereas a LUN is just a supporting element that contributes to how the virtual machine functions. By moving administration up to the VM level, policies can be applied more evenly across the infrastructure.

To VSA or Not to VSA?

There is a lot being written these days about why virtual storage appliances, orVSAs, (which run in user space) are terrible, why VSAs are awesome, why hypervisor converged (kernel space) storage management is terrible, and why hypervisor converged storage management is awesome. In short, should storage management services run in user space (VSA) or kernel space (kernel integrated)?

Defining VSA and Kernel-Integrated Management
Before examining the facts behind these opinions, let's take a minute to make sure you understand what constitutes a VSA versus a kernel-integrated storage management system. Bear in mind that both VSAs and kernel-integrated management systems are part of the software-defined storage family of storage systems in which storage resides in the server, not on SANs or separate arrays – at least in general.

A VSA is a virtual machine that runs on a host computer. This virtual machine's purpose is to manage the storage that is local to that host. The VSAs on individual hosts work together to create a shared storage pool and global namespace. This storage is then presented back to the virtual hosts and used to support virtual machines in the environment. Hyperconverged infrastructure companies, such as SimpliVity, Nutanix, and Maxta, all use VSAs to support the storage element of the solution.

Figure 2-1 provides a conceptual look at how VSAs operate. The key point here is to understand that the VSA is a virtual machine just like any other.

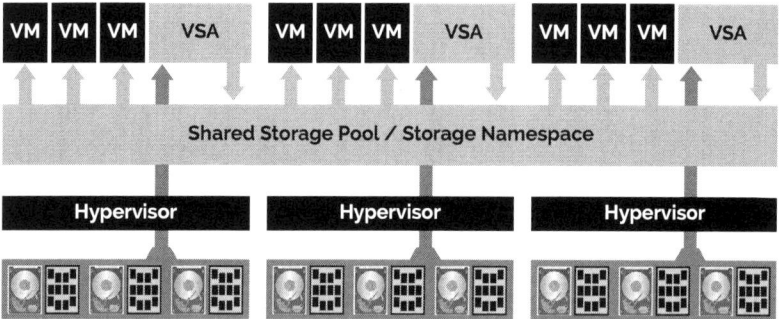

Figure 2-1: This is the general architecture that includes a VSA

Most hyperconverged systems on the market use this VSA method for handling storage abstraction.

However, kernel-integrated storage is another method you should understand. Referred to as kernel-integrated storage management or hypervisor-converged storage, this non-VSA storage management method operates through the implementation of a kernel-based module that resides in the hypervisor. In other words, instead of a virtual machine handling local storage management, this hypervisor kernel handles the job. The most well-known kernel-integrated hyperconverged infrastructure solutions are VMware VSAN/EVO:RAIL and Gridstore, which uses an operating system driver to handle storage needs.

Choosing a Method

So, which method is better? Let's take a look at both options and how they align with needs around hyperconverged infrastructure.

First, recall the discussion around hypervisor choice. If you don't need multi-hypervisor support, then either a VSA or a kernel-integrated kernel module will work equally well. Remember that multi-hypervisor choice is often not a legitimate requirement as long as the intended solution supports the hypervisor you want to use, or plan to use in the future. How do we know this? In our market report, ActualTech Media's *2015 State of Hyperconverged Infrastructure Market*, only 12% of respondents felt that multiple hy-pervisor support was a critical feature in a hyperconverged solution. Even then, only 26% of respondents felt that support for a specific hypervisor was important, leading one to believe that people would be willing to migrate to an alternative hypervisor if the hyperconverged infrastructure solution made sense.

As soon as you introduce a need for multi-hypervisor support, your only choice is to work with a VSA. Because a VSA is just another virtual machine running on the host, that VSA can be easily transitioned to run on any other hypervisor. When it comes to portability, VSA is king. There are far more VSA-based hyperconverged infrastructure solutions available on the market.

Hypervisor-integrated systems will lock you into the hypervisor to which the kernel module is tied. For some, that's a big downside. For others, it's not a problem since they don't have any plans or desire to move to a different hypervisor.

Finally, let's talk reality. VMware has spent years tuning the general hypervisor for performance and has told their customers that it's more than sufficient for running even their most performance-sensitive applications, including monster databases on Oracle and SQL Server, Exchange, and SAP. It boils down to this: If it is good enough for those kinds of really I/O-heavy applications, why can't it support storage and hyperconvergence?

It's hard to say that one solution is "better" than the other. Instead, they're just different ways to achieve the same goal, which is to abstract storage, pool it, and present it back to the hypervisor as a single shared resource pool. The choice really comes down to other goals you may have for your environment.

The Role of Custom Hardware in a Commodity Infrastructure

The first *S* in *SDS* stands for *software*. SDS is very much a software-driven storage architecture. However, this doesn't mean that custom hardware has no place in the solution. For software-defined purists, having any custom or proprietary hardware anywhere in the software-defined data center might be considered blasphemous. However, don't forget that we live in a world where not all is black and white. Shades of gray (no reference to the book intended!) permeate everything we do.

The purists are right to a point. Proprietary hardware that doesn't serve a strategic purpose doesn't belong in a software defined data center. However, when proprietary hardware provides a significant value-add that significantly differentiates a solution, it's worth a hard look. The vendor isn't creating that proprietary hardware for no reason.

SimpliVity is one vendor that has made the strategic decision to include a proprietary hardware card in their hyperconverged infrastructure solution. The company's accelerator card handles much of the heavy lifting when it comes to complex storage operations.

In the modern data center, some storage truths must be observed. The first is that latency is enemy number one. The more latency that is introduced into the equation, the slower that workloads will operate. SimpliVity's accelerator card is inserted into a commodity server and uses custom-engineered chips to provide ultra-fast write caching services that don't rely on commodity CPUs. Moreover, the accelerator card enables comprehensive data reduction technologies (deduplication and compression) to take place in real-time with no performance penalty in order to massively reduce the total amount of data that has to be stored to disk and the I/O that it takes to carry out operations.

Even when there is some custom hardware, "software defined" has nothing to do with hardware. Software defined is about abstraction from the underlying hardware, thereby allowing software to present all services to applications.

Decision 3: Data Protection Services
Data protection shouldn't be considered an afterthought in your data center. It should be considered a core service that is central to how IT operates. Recovery Time Objectives (RTOs) and Recovery Point Objectives (RPOs) should be a key discussion point as you're considering hyperconverged infrastructure solutions. Bear in mind that not all hyperconverged products come with the same levels of data protection.

Decision 4: The Management Layer

The data center has become an ugly place when it comes to management. There are separate administrative consoles for everything in the environment. The result is that administrators have no consistency in their work and are burdened with inefficiencies. To simplify management in the data center, admins need as few interfaces as possible. Here are the most common options that you need to be aware of when considering a hyperconverged virtual infrastructure:

Virtualization Layer Management

For those using VMware vSphere, vCenter is going to be the virtualization layer management tool that must be in place. Organizations using Microsoft Hyper-V will use System Center Virtual Machine Manager (SCVMM).

Orchestration and Automation Layer Management

Once the hyperconverged infrastructure is running, common tasks must be automated to gain efficiencies. Common orchestration and automation tools are:

- **VMware's vRealize Automation (vRA)** — Provides automated provisioning through a service catalog. With the ability to deploy across multi-vendor cloud and virtual infrastructures, vRA allows you to provide the applications to the business as needed.

- **Cisco's Unified Computing System Director (or UCSD)** — For those using Cisco UCS Servers, UCSD offers dynamic provisioning, dynamic hardware, and significant reduction in management points.

- **OpenStack** — Rapidly growing in popularity, OpenStack is being adopted by enterprises to create private cloud infrastructures. OpenStack is highly customizable and offers the lowest entry cost of any cloud management platform (CMP) because of its open source price tag. KVM is the default hypervisor for OpenStack, therefore your main concern may be whether or not the HCI solution you choose supports multiple hypervisors, and specifically KVM.

Vendor-Supplied Management

Many hyperconvergence solutions provide you with a whole new management interface for your virtual infrastructure. With most hyperconverged solutions running vSphere today, this idea of creating a whole new management tool for the virtual infrastructure disregards that fact that you already have one — VMware vCenter.

REST APIs

A Relative State Transition Application Programming Interface (REST API) provides you with the entry point required to integrate multiple management points and automate your data center.

You should ensure that the hyperconvergence solution you choose offers compatibility with the virtualization management, automation, orchestration, and API tools discussed here. Also, ensure that your hyperconvergence solution does whatever is possible to reduce the number of management points and tools that are required for administration and troubleshooting.

Hypervisor Support

By its very nature, hyperconverged infrastructure requires using some kind of hypervisor. The hypervisor has become the standard layer on which most new business applications are deployed. Although there are still services deployed on bare metal servers, they are becoming far less common as virtualization assimilates more and bigger workloads.

With virtualization forming the core for hyperconvergence infrastructure solutions, the question naturally turns to one of hypervisor choice. If there's one thing IT administrators try to avoid, it's a lack of choice. Organizations demand choice, and this is also true when considering the server virtualization component of the data center.

Just keep in mind a few key facts when choosing hypervisor support:

- First, although variety in choice is highly desired, it's not always required for individual hyperconverged infrastructure solutions. There are options on the market today that each support vSphere, KVM, and Hyper-V. If you absolutely demand to be able to use a particular hypervisor, there is likely a solution waiting for you. However, not every hyperconvergence vendor supports every hypervisor.

- Second, for the hyperconverged infrastructure vendors that do support multiple hypervisors, the customer (that's you!) gets to decide which hypervisor to run on that platform. With that said, we discovered in our 2015 State of Hyperconverged Infrastructure Market report that people don't believe that supporting multiple hypervisors is that important.

In terms of which hypervisors are the most popular, it's pretty common knowledge that the most popular hypervisor on the market today is VMware's vSphere. With a formidable command of the hypervisor market, vSphere is offered as a primary hypervisor choice on a number of hyperconverged platforms.

Therefore, your decision around hypervisor support is really simple: For which hypervisor do you require support?

Up Next

Overall data center architecture includes a number of critical decision points, as described in this chapter. You were briefly introduced to one decision point: hardware acceleration. However, there's a lot more to that story. You will learn a lot more in Chapter 3 about the additional benefits provided by hardware acceleration.

3

Addressing Data Center Metrics Pain Points

As much as IT pros hate to be told, "We have to do more with less," it's doubtful that this directive will die in the near future. The unfortunate truth is that IT has to continue to do more with either no increase or with decreases in overall resources. This comes at the same time that increasing attention must be paid to various traditional data center pain points.

In this chapter, you will learn about these pain points and how hyperconverged infrastructure can be leveraged to help address them.

The Relationship Between Performance & Virtual Machine Density

Return on investment. Total cost of ownership. These are phrases used to describe the economic impact of technology investments — or expenses — depending on your perspective. Regardless of the perspective though, businesses want to squeeze as much return as possible out of their technology investments while spending as little as reasonably possible on those investments.

You might be wondering what this quick economic discussion has to do with workload performance in the data center. There is actually a direct link between these two topics and it revolves around overall virtual machine density. Virtual machine density refers to the number of virtual machines that you can cram onto a single host. The more virtual machines that you can fit onto a host, the fewer hosts you need to buy. Obviously, fewer hosts means having to spend less money on hardware, but the potential savings go far beyond that measure.

When you have fewer hosts, you also spend less on licensing. For example, you don't need to buy vSphere licenses for hosts that don't exist! In addition, if you've taken Microsoft up on their Windows Server Data Center licensing deal under which you simply license virtual hosts, and you can run as many individual virtual machine-based Windows instances as you like under that Data Center umbrella, you save even more.

The savings don't stop there. Fewer hosts means less electricity is needed to operate the data center environment. Fewer hosts means there is less cooling needed in the data center environment. Fewer hosts means that you free up rack space in the data center environment.

However, these benefits cannot come at the expense of poor workload performance. When workloads perform poorly, they actively cost the company money, such as lost efficiency and customer dissatisfaction.

How do you maximize virtual machine density without impacting workload performance? First of all, it's a balance that you need to find, but when you're initially specifying hardware for a new environment, you won't necessarily know how your workloads will function in that new environment, so things can be tough to predict. Instead, you need to look at the inputs, or the resources atop which the new environment is built. Storage is one of these key resources.

Storage Performance in a Hyperconverged Infrastructure

In a hyperconverged infrastructure environment, one of the primary resources that must be considered is storage, and not just from a capacity perspective. Remember that storage and compute are combined in hyperconvergence, so that becomes a factor that is not present in more traditional environments. In a traditional environment, 100% of the available CPU and memory resources are dedicated to serving the needs of running virtual machines. In a hyperconverged infrastructure environment, some of those resources must be diverted to support the needs of the storage management function, usually in the form of a VSA. This is one of the core trade-offs to consider when adopting a hyperconverged infrastructure.

This is where hardware acceleration can be a boon. Most hyperconverged infrastructure systems rely on the commodity hardware to carry out all functions. With a system that uses hardware acceleration, more commodity Intel CPU horsepower can be directed at running virtual machines while the acceleration hardware handles processor-intensive data reduction operations, such as deduplication and compression.

Data Deduplication Explained

Consider this scenario: Your organization is running a virtual desktop environment with hundreds of identical workstations all stored on an expensive storage array purchased specifically to support this initiative. That means you're running hundreds of copies of Windows, Office, ERP software, and any other tools that users require.

Let's say that each workstation image consumes 25 GB of disk space. With just 200 such workstations, these images alone would consume 5 TB of capacity.

With deduplication, you can store just 1 copy of these individual virtual machines and then allow the storage array to place pointers to the rest. Each time the deduplication engine comes across a piece of data that is already stored somewhere in the environment, rather than write that full copy of data all over again, the system instead saves a small pointer in the data copy's place, thus freeing up the blocks that would have otherwise been occupied.

In the figure "Deduplication vs. No Deduplication", the graphic on the left shows what happens without deduplication. The graphic on the right shows deduplication in action. In this example, there are four copies of the **gray** block and two copies of the **black** block stored on this array. Deduplication enables just one block to be written for each block, thus freeing up those other four blocks.

Now, let's expand this example to a real-world environment. Imagine the deduplication possibilities present in a VDI scenario: with hundreds of identical or close-to-identical desktop images, deduplication has the potential to significantly reduce the capacity needed to store all of those virtual machines.

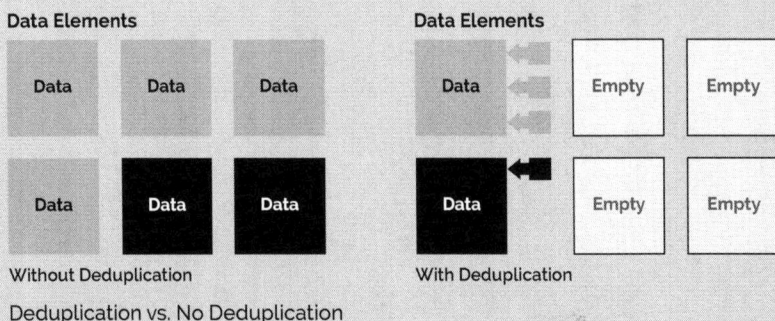

Data Elements

Data	Data	Data
Data	Data	Data

Without Deduplication

Data Elements

Data	Empty	Empty
Data	Empty	Empty

With Deduplication

Deduplication vs. No Deduplication

Deduplication works by creating a data fingerprint for each object that is written to the storage array. As new data is written to the array, additional data copies beyond the first are saved as tiny pointers. If a completely new data item is written — one that the array has not seen before — the full copy of the data is stored.

Different vendors handle deduplication in different ways. In fact, there are two primary deduplication techniques that deserve discussion: inline deduplication and post-process deduplication.

Inline Deduplication

Inline deduplication takes place at the moment in which data is written to the storage device. While the data is in transit, the deduplication engine fingerprints the data on the fly. As you might expect, this deduplication process does create some overhead.

First, the system has to constantly fingerprint incoming data and then quickly identify whether that new fingerprint already matches something in the system. If it does, a pointer to the existing fingerprint is written. If it does not, the block is saved as-is. This process introduces the need to have processors that can keep up with what might be a tremendous workload. Further, there is the possibility that latency could be introduced into the storage I/O stream due to this process.

A few years ago, this might have been a showstopper since some storage controllers may not have been able to keep up with the workload need. Today, though, processors have evolved far beyond what they were just a few years ago. These kinds of workloads don't have the same negative performance impact that they might have once had. In fact, inline deduplication is a cornerstone feature for most of the new storage devices released in the past few years and, while it may introduce some overhead, it often provides far more benefits than costs. With a hardware-accelerated hyperconverged infrastructure, inline deduplication is not only the norm, it's a key cornerstone for the value that is derived from the infrastructure.

Post-Process Deduplication

As mentioned, inline deduplication imposes the potential for some processing overhead and potential latency. The problem with some deduplication engines is that they have to run constantly, which means that the system needs to be adequately configured with constant deduplication in mind. Making matters worse, it can be difficult to predict exactly how much processing power will be needed to achieve the deduplication goal. As such, it's not always possible to perfectly plan overhead requirements.

This is where post-process deduplication comes into play. Whereas inline deduplication processes deduplication entries as the data flows through the storage controllers, post-process deduplication happens on a regular schedule, perhaps overnight. With post-process deduplication, all data is written in its full form — copies and all — on that regular schedule. The system then fingerprints all new data and removes multiple copies, replacing them with pointers to the original copy of the data.

Post-process deduplication enables organizations to utilize this data reduction service without having to worry about the constant processing overhead involved with inline deduplication. This process enables organizations to schedule dedupe (deduplication) to take place during off hours.

The biggest downside to post-process deduplication is the fact that all data is stored fully hydrated – a technical term that means that the data has not been deduplicated – and, as such, requires all of the space that non-deduplicated data needs. It's only after the scheduled process that the data is shrunk. For those using post-process dedupe, bear in mind that, at least temporarily, you'll need to plan on having extra capacity. There are a number of hyperconverged infrastructure systems that use post process deduplication while others don't do deduplication at all. Lack of full inline deduplication increases costs and reduces efficiency.

Hardware Acceleration to the Rescue

Hardware-accelerated hyperconverged infrastructure solutions completely solve the overhead challenges inherent in those systems. All deduplication tasks are delegated to the accelerator, thereby negating the need for the system to consume processor resources that are also needed by the virtual machines.

Tiering and Deduplication

In order to match storage performance needs with storage solutions, many companies turn to what are known as tiered storage solutions. They run, for example, hard disk-based arrays for archival data, and they run flash system for performance needs, and they manage these resources separately. This also means that deduplication is handled separately per tier. Each time dedupe is duplicated, there are additional CPU resources that must be brought to bear and there are multiple copies of deduplicated data. Neither is efficient. Hyperconverged systems that include comprehensive in-line deduplication services carry with them incredibly efficient outcomes.

Don't underestimate the benefits of data reduction! These services have far more impact on the environment than might be obvious at first glance and the benefits go far beyond simple capacity gains, although capacity efficiency is important.

There are several metrics that benefit when dedicated and specialized hardware is brought to bear.

Capacity

The sidebar, **Data Deduplication Explained**, discusses generalized capacity benefits of deduplication, but let's now consider this in the world of hyperconverged infrastructure. In order to do this, you need to consider your organization's holistic data needs:

- **Primary storage** — This is the storage that's user- or application-facing. It's where your users store their stuff, where email servers store your messages, and where your ERPs database is housed. It's the lifeblood for your day-to-day business operations.

- **Backup** — An important pillar of the storage environment revolves around storage needs related to backup. As the primary storage environment grows, more storage has to be added to the backup environment, too.

- **Disaster recovery** — For companies that have disaster recovery systems in place in which data is replicated to secondary data centers, there is continued need to grow disaster recovery-focused storage systems.

When people think about storage, they often focus just on primary storage, especially as users and applications demand more capacity. But when you look at the storage environment from the top down, storage growth happens across all of these storage tiers, not just the primary storage environment.

In other words, your capacity needs are growing far faster than it might appear. Hardware acceleration, when applied to all of the storage domains in aggregate, can have a tremendous impact on capacity. By treating all of these individual domains as one, and deduplicating across all of them, you can achieve pretty big capacity savings.

But deduplication, as mentioned before, can be CPU-intensive. By leveraging hardware acceleration, you can deduplicate all of this without taking CPU resources away from running workloads. By infusing the entire storage environment with global deduplication capabilities via hardware acceleration, you can get capacity benefits that were only the stuff of dreams just a few years ago. Hyperconvergence with great deduplication technology can attain great results while also simplifying the overall management needs in the data center.

IOPS

Imagine a world in which you don't actually have to write 75% of the data that is injected into a system. That world becomes possible when hardware acceleration is used so that all workloads benefit from inline deduplication, not just some workloads.

The more that data can be deduplicated, the fewer write operations that have to take place. For example, a deduplication ratio of 5:1 means that there would only be 1 actual write-to-storage operation that takes place for every 5 attempted write operations.

Hardware acceleration allows this comprehensive data reduction process to take place in a way that doesn't divert workload

CPU resources. As a result, you continue to enjoy the IOPS benefits without having to give up workload density.

Latency

Latency is the enemy of data center performance. By offloading intensive tasks to a custom developed hardware board that specifically handles these kinds of tasks, latency can be reduced to a point where it doesn't affect application performance.

Application Performance

At the end of the day, all that matters is application performance. That's the primary need in the data center environment and, while it can be difficult to measure, you will know very quickly if you've failed to hit this metric. The phones will start to ring off the hook. Hardware acceleration helps you to keep this metric in the green.

Linear Scalability

Businesses grow all the time. The data center has to grow along with it. Scaling "up" has been one of the primary accepted ways to grow, but it carries some risks. Remember, in the world of storage, scaling up occurs when you add additional capacity without also adding more CPU and networking capacity at the same time. The problem here is that you run the risk of eventually overwhelming the shared resources that exist. **Figure 3-1** shows an example of a scale-up environment.

Scale-out has become a more popular option because it expands all resources at the same time. With hyperconverged infrastructure, the scaling method is referred to as *linear scalability*. Each node has all of the resources it needs — CPU, RAM, and storage — in order to stand alone. **Figure 3-2** gives you a look at this kind of scalability.

Scale Up Architecture - Increasing Burden on Shared Components and Single Points of Failure

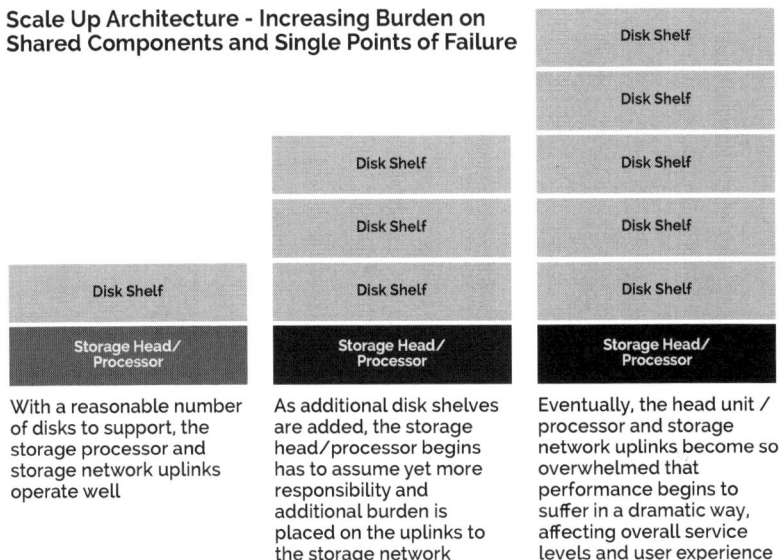

		Disk Shelf
		Disk Shelf
	Disk Shelf	Disk Shelf
	Disk Shelf	Disk Shelf
Disk Shelf	Disk Shelf	Disk Shelf
Storage Head/ Processor	Storage Head/ Processor	Storage Head/ Processor

With a reasonable number of disks to support, the storage processor and storage network uplinks operate well	As additional disk shelves are added, the storage head/processor begins has to assume yet more responsibility and additional burden is placed on the uplinks to the storage network	Eventually, the head unit / processor and storage network uplinks become so overwhelmed that performance begins to suffer in a dramatic way, affecting overall service levels and user experience

Figure 3-1: A scale up environment relies on shared components

Horizontal Scale Architecture - Shared Nothing and Linear Resource Scalability

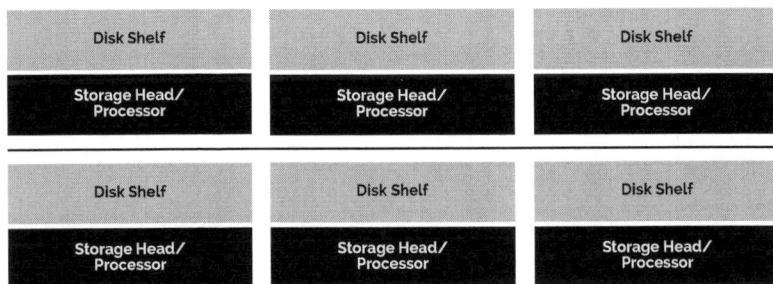

Disk Shelf	Disk Shelf	Disk Shelf
Storage Head/ Processor	Storage Head/ Processor	Storage Head/ Processor
Disk Shelf	Disk Shelf	Disk Shelf
Storage Head/ Processor	Storage Head/ Processor	Storage Head/ Processor

Figure 3-2: A scale out environment has nodes that can individually stand alone

For solutions that use hardware acceleration, accelerators are a critical part of the scaling capabilities as they offload intensive functionality that can be workload impacting. This increases density, but more importantly, by freeing up resources, IT can add more predictability to overall performance of applications, even while maintaining high levels of density.

Up Next

The items discussed in this chapter are critically important to ensuring that the data center adequately (maybe even excellently!) supports business needs, but these metrics are just the very tip of the iceberg. Under the waterline are other requirements that can't be ignored. In the next chapter, we'll discuss one really important requirement that is often quite challenging: data protection.

4

Ensuring Availability, Data Protection and Disaster Recovery

Even the smallest of small businesses today depend on their IT resources being available on a 24/7 basis. Even short periods of downtime can wreak havoc, impact the bottom line, and mean having to cancel going out to lunch. Maintaining an agreed-upon level of infrastructure availability is critically important. On top of that, outages or other events resulting in loss of data can be a death knell for the business. Many businesses that suffer major data loss fail to recover in the long term and eventually make their way down the drain. Data protection is one of IT's core services. Unfortunately, it's also a hard service to provide at times, or at least, it was. There are now some hyperconverged infrastructure solutions that are uniquely positioned to solve, once and for all, the challenges across the entire data protection spectrum.

The Ins & Outs of Backup & Recovery

There are two primary metrics to consider when it comes to disaster recovery.

Recovery Point Objective (RPO)

If you're using a nightly backup system, you're implicitly adhering to a *24-hour Recovery Point Objective (RPO)*. You're basically saying that losing up to 24 hours worth of data is acceptable to the business. RPO is the metric that defines how much data your organization is willing to lose in the event of a failure that has the potential to result in data loss. To reduce RPO, you need to back data up more often.

Recovery Time Objective (RTO)

RPO is critically important as it defines just how much data you're willing to lose. Once you've suffered a data loss, the critical metric shifts. Now, you're more interested in how long it takes you to recover from that failure. How long is your organization willing to be without data while you work to recover it from backup systems? This metric is often used to support such statements as, "For every minute we're down, the company loses $X."

The Recovery Time Objective (RTO) is the formal name for this metric and is one that companies will go to great lengths to minimize. As is the case with RPO, the closer to zero that you attempt to get to RTO — that is, the less time that you're willing to be down — the more it costs to support.

To achieve very low RTO values, companies will often implement multi-pronged solutions, such as disaster recovery sites, fault tolerant virtual machines, clustered systems, and more.

The Data Protection and Disaster Recovery Spectrum

Let's talk a bit about data protection as a whole. When you really look at it, data protection is a spectrum of features and services. If you assume that data protection means "ensuring that data is available when it's needed," the spectrum also includes high availability for individual workloads. **Figure 4-1** provides you with a look at this spectrum.

Figure 4-1: The Data Protection Spectrum

RAID

Yes, RAID is a part of your availability strategy, but it's also a big part of your data protection strategy. IT pros have been using RAID for decades. For the most part, it has proven to be very reliable and has enabled companies to deploy servers without much fear of negative consequences in the event of a hard drive or two failing. Over the years, companies have changed their default RAID levels as the business needs have changed, but the fact is that RAID remains a key component in even the most modern arrays.

The RAID level you choose is really important, but you shouldn't have to worry about it. The solution should do it for you. That said, don't forget that it's pretty well-known that today's really large hard drives have made traditional RAID systems really tough to support. When drives fail in a traditional RAID array, it can take hours or even days to fully rebuild that drive. Don't forget this as you read on; we'll be back to this shortly.

RAID is also leveraged in some hyperconverged infrastructure systems; however, with these systems, administrators are shielded from some of the complexity and configuration options that they used to work with on stand-alone storage arrays. Bear in mind that one of the tenets of hyperconverged infrastructure is simplicity. As such, you don't have to go through a lot of effort to manage RAID in a hyperconverged system. It's simply leveraged behind the scenes by the system itself. In **Figure 4-2**, you get a look at how RAID protects data.

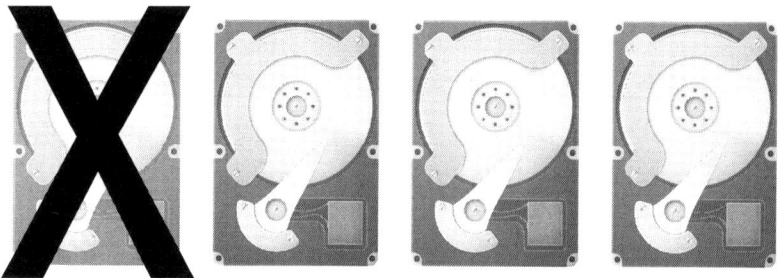

Figure 4-2: Key Takeaway: On the data protection spectrum, RAID helps you survive the loss of a drive or two.

Replication/RAIN/Disaster Recovery

RAID means you can lose a drive and still continue to operate, but what happens if you happen to lose an entire node in a hyperconverged infrastructure cluster? That's where replication jumps in to save the day. Many hyperconverged infrastructure solutions on the market leverage replication as a method for ensuring ongoing availability and data protection in the event that something takes down a node, such as a hardware failure or an administrator accidentally pulling the wrong power cord.

This is possible because *replication* means "making multiple copies of data and storing them on different nodes in the cluster." Therefore, if a node is wiped off the face of the earth, there are one or more copies of that data stored on other cluster nodes.

Two kinds of replication

There are two different kinds of replication to keep in mind. One is called *local* and the other is called *remote*. Local replication generally serves to maintain availability in the event of a hardware failure. Data is replicated to other nodes in the cluster in the same data center. Remote replication is leveraged in more robust disaster recovery scenarios and enables organizations to withstand the loss of an entire site.

In some hyperconverged infrastructure solutions, like those shown in **Figure 4-3**, you can configure what is known as the *replication factor* (RF). The replication factor is just a fancy way of telling the system how many copies of your data you'd like to have. For example, if you specify a replication factor of 3 (RF3), there will be 3 copies of your data created and stored across disparate nodes. You will sometimes see replication-based availability mechanisms referred to as RAIN, which stands for Redundant Arrays of Independent Nodes.

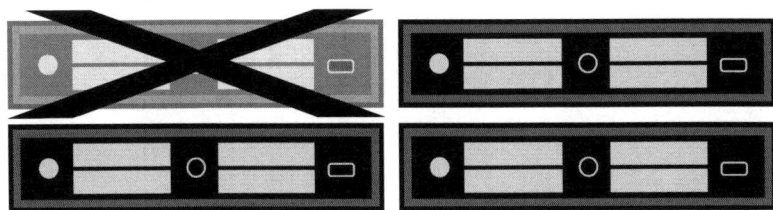

Figure 4-3: Lost a node? Can't find it? Don't worry! Replication will save the day!

Besides helping you to make sure that your services remain available, replication goes way beyond just allowing you to withstand the loss of a node, too. When replication is taken beyond the data center to other sites, you suddenly gain disaster recovery capability, too. In fact, in some hyperconverged systems that leverage inline deduplication across primary and secondary storage tiers, that's exactly what happens. After deduplication, data is replicated to other nodes and to other data centers, forming the basis for incredibly efficient availability and disaster recovery.

How About Both – RAID and RAIN Combined

Let's go a little deeper into the RAID/RAIN discussion with an eye on hyperconverged infrastructure solutions that provide both. First, there are some downsides to just RAIN-based replication (Replication Factor 2 or RF2). There are solutions on the market that provide RF2. Systems based on RF2 will lose data if any two nodes or disks in a cluster fail, or if even just one node should fail while any other node is down for maintenance.

To make things a bit more resilient, you could bump up to RF3, but this replication factor then requires a minimum of five nodes at each site that uses RF3 and imposes an additional 50% penalty on capacity. With RF3, you can also start to think about using erasure coding, but this requires RF3 and carries with it a lot of CPU overhead due to the way that erasure coding works. This may not be suitable when trying to support high-performance applications.

How about combining RAID and RAIN into a single solution? Maybe you combine the use of local RAID 6 on individual nodes so that *any* node can tolerate double disk failures and keep virtual machines up and running. With each individual node very well protected, the likelihood of losing an entire node is reduced. From there, you apply RAIN so that, in the event that a complete node is lost, you can tolerate that, too. The strategic combination of RAID and RAIN enables tolerance against a broad set of failure scenarios.

What is Erasure Coding?

Erasure coding is usually specified in an N+M format: 10+6, a common choice, means that data and erasure codes are spread over 16 (N+M) drives, and that any 10 of those can recover data. That means any six drives can fail. If the drives are on different appliances, the protection includes appliance failures, so six appliance boxes could go down without stopping operations.

Courtesy: www.networkcomputing.com/storage/ raid-vs-erasure-coding/a/d-id/1297229

Backup and Recovery

Despite your best efforts, there will probably come a day when you need to recover data lost from production. Data losses can happen for a variety of reasons:

- **Human error** — People make mistakes. Users accidentally delete files. Administrators accidently delete virtual machines. IT pros can sometimes accidentally pull the wrong disk from a storage system or unplug the wrong server's power cord.

- **Hardware failure** — When hardware fails, sometimes it fails spectacularly. In fact, hardware failure may not even be the result of failed IT hardware. You may end up in a situation, for example, in which the data center cooling systems fail and server automatically shuts down as the temperature rises. This could be considered a server hardware failure because of the outcome (the server going down), when in fact the server is actually doing exactly what it's supposed to do in this case.

- **Disasters** — Hurricanes, tornados, floods, a new Terminator movie. Disasters come in all kinds of forms and can result in data loss.

The SimpliVity story on protecting production data and availability
by Brian Knudston

Being a hyperconvergence platform, SimpliVity first provides the compute and storage infrastructure for customer's production applications. As data is ingested from the hypervisor, we stage the VM (virtual machine) data into DRAM on the OmniStack Accelerator Card across two of our nodes within a single data center. With data now protected across multiple nodes, in addition to supercapacitor and flash storage protecting the DRAM on each OmniStack Accelerator Card, we acknowledge a successful write back to the VM and process the data for deduplication, compression and optimization to permanent storage on the Hard Disk Drives (HDDs) on both nodes. Once this process is complete, every VM in a SimpliVity data center can survive the loss of at least two HDDs in every node, in a data center AND the loss of a full SimpliVity node.

Disaster Recovery

Disaster recovery takes backup one step further than the basics. Whereas *backup* and *recovery* are terms that generally refer to backing up data and, when something happens, recovering that data, *disaster recovery* instead focuses on recovery beyond just the data.

Disaster recovery demands that you think about the eventual needs by starting at the end and working backward. For example, if your data center is hit by an errant meteor (and assuming that this meteor has not also caused the extinction of the human race) recovering your data alone will be insufficient. You won't have anything onto which to recover your data if your data center is obliterated.

Before we get too fatalistic, let's understand what the word *disaster* really means in the context of the data center. It's actually kind of an unfortunate term since it immediately brings to mind extinction-level events, but this is not always the case for disaster recovery.

There are really two kinds of disasters on which you need to focus:

- **Micro-level disasters** — These are the kinds of events that are relatively common, such as losing a server or portion of a data center. In general, you can quickly recover in the same data center and keep on processing. Often, recovery from these kinds of disasters can be achieved through backup and recovery tools. With that said, these events will probably still result in downtime.

- **Macro-level disasters** — These are the kind of life-altering events that keep IT pros awake at night and include things like fires, acts of {insert deity here}, or rampaging hippos. Recovery from these disasters will mean much more than just restoring data.

Business Continuity

Since *disaster recovery* is kind of a loaded term, a lot of people prefer to think about the disaster recovery process as "business continuity" instead. However, that's not all that accurate. Business continuity is about all the aspects to a business continuing after a disaster. For example, where are the tellers going to report after the fire? How are the phone lines going to be routed? Disaster recovery is an IT plan that is a part of business continuity.

Thinking about the disaster recovery process with the end in mind requires that you think about what it would take to have everything back up and running — hardware, software, and data — before disaster strikes.

Yes, your existing backup and recovery tools probably play a big role in your disaster recovery plan, but that's only the very beginning of the process.

Disaster recovery plans also need to include, at a bare minimum:

- **Alternate physical locations** — If your primary site is gone, you need to have other locations at which your people can work.

- **Secondary data centers** — In these locations, or in the cloud, you need to have data centers that can handle the designated workloads from the original site. This includes a space for the hardware, the hardware itself, and all of the software necessary to run the workloads.

- **Ongoing replication** — In some way, the data from your primary site needs to make its way to your secondary site.

This is a process that needs to happen as often as possible in order to achieve desirable RTOs and RPOs. In an ideal world, you would have systems in place that can replicate data in minutes after it has been handled in the primary data center. The right hyperconverged infrastructure solution can help you to achieve these time goals.

- **Post-disaster recovery processes** — Getting a virtual machine back up and running is just the very first step in a disaster recovery process. RTO is a measure of more than just the restoration of the VM. From there, processes need to kick off that include all the steps required to get the application and data available to the end user. These processes include IP address changes, DNS updates, re-establishment of communication paths between parts of an n-tier application stack and other non-infrastructure items.

SimpliVity's answer to full spectrum DR

by Brian Knudston

SimpliVity alone makes it simple for you to achieve the first part of disaster recovery, which is making sure that virtual machines are always available, even if a data center is lost. SimpliVity has made a focus over the last several months to provide integration into other tools that can help automate and orchestrate all of the remaining steps of the disaster recovery process, including pre-built packages of SimpliVity functionality within VMware's vRealize Automation and Cisco's UCS Director, and supporting partners in the development of tools on top of SimpliVity APIs like VM2020's EZ-DR.

Data Reduction in the World of Data Protection

We're going to be talking a lot about data reduction – deduplication and compression – in this book. They're a huge part of the hyperconverged infrastructure value proposition and, when done right, can help IT to address problems far more comprehensively than when it's done piecemeal.

When it comes to data protection, data reduction can be really important, especially if that data reduction survives across different time periods – production, backup, and disaster recovery. If that data can stay reduced and deduplicated, some very cool possibilities emerge. The sidebar below highlights one such solution.

The Data Virtualization Platform and disaster recovery

by Brian Knudston

To protect data at specific instances of time, SimpliVity designed backup and restoration operations directly into the DNA of the SimpliVity OmniStack Data Virtualization Platform, enabled by our ability to dedupe, compress and optimize all the VM data. This results in backups and restores that can be taken in seconds, which can help reduce Recovery Point Objectives (RPOs) and Recovery Time Objectives (RTOs), while consuming almost no IOPS off the HDDs.

When protecting data across data centers, SimpliVity maintains awareness of data deduplication across the different sites. If a VM is configured to back up to a remote data center, the receiving data center determines which unique blocks need to be transported across the WAN and the sending data center only sends those unique blocks. This drastically reduces the WAN bandwidth necessary between sites, increasing the frequency of backups to remote sites and eliminate IOPS by reducing the amount of data that needs to be read from and written to the HDDs.

Fault Tolerance

Availability is very much a direct result of the kinds of fault tolerance features built into the infrastructure as a whole. Data center administrators have traditionally taken a lot of steps to achieve availability, with each step intended to reduce the risk of a fault in various areas of the infrastructure. These can include:

- **Using RAID** — As previously mentioned, RAID allows you to experience drive failures within a hyperconverged node and keep operating.

Simplified Storage Systems

Bear in mind that RAID, and storage in general, becomes far simpler to manage in a hyperconverged infrastructure scenario. There is no more SAN and, in most cases, RAID configuration is an "under the hood" element that you don't need to worry about. This is one less component that you have to worry about in your data center.

- **Redundant power supplies** — Extra power supplies are, indeed, a part of your availability strategy, because they allow you to experience a fault with your power system and still keep servers operating.

- **Multiple network adapters** — Even network devices can fail, and when they do, communications between servers and users and between servers and other servers can be lost. Unless you have deployed multiple switches into your environment and multiple network adapters into your servers, you can't survive a network fault. Network redundancy helps you make your environment resilient to network-related outages.

- **Virtualization layer** — The virtualization layer includes its own fault tolerance mechanisms, some of which are transparent and others require a quick reboot. For example, VMware's High Availability (HA) service continuously monitors all of your vSphere hosts. If one fails, workloads are automatically restarted on another node. There is some downtime, but it's minimal. In addition to HA, VMware makes available a Fault Tolerance (FT) feature. With FT, you actually run multiple virtual machines. One is the production system and the second is a live shadow VM that springs into action in the event that the production system becomes unavailable. However, with all that said, there are some limitations inherent in hypervisor-based fault tolerance technology, described in the sidebar entitled *Fault Tolerance Improvements in vSphere 6.* This is why some hyperconverged infrastructure vendors eschew hypervisor-based fault tolerance mechanisms in favor of building their own, more robust solutions.

Fault Tolerance Improvements in vSphere 6

Frankly, Fault Tolerance (FT) in vSphere has been all but useless, except for the smallest virtual machines. Here's an excerpt from VMware's documentation explaining the limitations of FT: "Only virtual machines with a single vCPU are compatible with Fault Tolerance." This limitation is one of the many items that holds back FT from being truly usable across the board. vSphere 6 increases Fault Tolerance capabilities to virtual machines with up to 4 vCPUs. This is still a significant limitation when you consider than many VMs are deployed with 8 vCPUs or more, particularly for large workloads.

End Results:
High Availability, Architectural Resiliency, Data Protection, and Disaster Recovery

No one wants downtime. It's expensive and stressful. Most organizations would be thrilled if IT could guarantee that there would be no more downtime ever again. Of course, there is no way to absolutely guarantee that availability will always be 100%, but organizations do strive to hit 99% to 99.999% availability as much as possible.

High availability is really the result of a combination of capabilities in an environment. In order to enable a highly-available application environment, you need to have individual nodes that can continue to work even if certain hardware components fail and you need to have a cluster that can continue to operate even if one of the member nodes bites it.

Hyperconverged infrastructure helps you to achieve your availability and data protection goals in a number of different ways. First, the linear scale-out nature of hyperconverged infrastructure (i.e., as you add nodes, you add all resources, including compute, storage, and RAM), means that you can withstand the loss of a node because data is replicated across multiple nodes with RAIN. Plus, for some hyperconverged solutions, internal use of RAID means that you can withstand the loss of a drive or two in a single node. With the combination of RAIN+RAID providing the most comprehensive disaster recovery capabilities, you can withstand the loss of an entire data center and keep on operating with little to no loss of data.

As you research hyperconverged infrastructure solutions, it's important to make sure that you ask a lot of questions about how vendors provide availability and data protection in their products. The answers to these questions will make or break your purchase.

Up Next

It's clear that data protection and high availability are key components in any data center today. The cloud has become another way that companies can improve their availability and data protection systems. Of course, cloud can do a lot more as well, which is why it's the topic of our next chapter.

5

Hyperconvergence & the Public Cloud

This chapter will help you understand the ways by which you can leverage cloud services as a part of your hyperconverged infrastructure solutions. It will also help you better understand the concept of "private cloud" and how that fits with hyperconvergence. Can a hyperconverged solution deliver some of the things that cloud can give you?

Why Is Cloud so Desirable?

You will learn more about what defines *cloud* a little later in this chapter. Before getting into the various definitions, though, let's discuss the traits inherent in cloud systems which make them a popular and desirable choice for service deployment.

The Economic Model

Everything eventually comes down to money. Business decision-makers are constantly on the lookout for ways to reduce costs while also boosting effi ciency and outcomes. This is often a seemingly impossible task described as "doing more with less." IT was supposed to be an enabler, but for many companies, it has become a money pit — an expense center to be minimized. Obviously, when leveraged properly, IT can be an incredible enabling function, but even in these cases, no one wants to spend more than they have to.

When you buy your own data center hardware and software, you incur pretty significant CapEx. This initial cash outlay necessary to procure a solution can be pretty high and can result in the need to cut corners or even delay upgrades if there is not enough cash available.

When you decide to start consuming resources from the public cloud, there is no initial cash outlay necessary. You don't incur capital expenses. Sure, you might have to pay a bit in the way of startup costs, but you don't have to buy hardware and software. You simply rent space on someone else's servers and storage.

Business decision-makers love this shift. They don't need to worry about huge capital outlays, and they know that they're paying for what they use. They're not paying for extra hardware that may never end up actually being leveraged to help solve business needs.

Scale

When you build your own data center, you have to scale it yourself. Sometimes, you can scale in increments that make financial sense, while other times you have to add more than you might like due to predefined requirements from your vendors.

When you use the public cloud, you don't have to worry about inherent scaling limits or increments. Remember, you pay for what you use. As your usage grows, so does your bill, but you don't generally need to manually add new resources to your account. It can happen automatically.

Scalability granularity often isn't a problem with the public cloud. You grow as you need to. There is no practical limit to how far you can grow as long as the cloud provider still has resources.

Geographic Diversity and Disaster Recovery

Building multiple data centers can be an expensive undertaking, but it's one that is being executed more and more as companies seek ways to protect their data and ensure continuity of their business in the event of a disaster striking the primary data center. The separate data centers are generally geographically diverse so that a single natural disaster can't take out both sites at the same time.

Public cloud providers often already have systems that can quickly enable geographic diversity for applications that are already running on their systems. Enabling geographic diversity is often as simple as clicking a mouse button and, most likely, paying some additional money to the cloud provider.

The Public Cloud

It's hard to avoid the term *cloud* today. It's everywhere. For many, the term itself has become synonymous with "Internet" or is just another way to describe what used to be called "hosted services." However, there are a number of traits that make a public cloud a public cloud.

First, in general, public cloud systems are comprised of multi-tenant environments operated by a service provider with the hardware and software located in the provider's data center. In these environments, the customer may not always even be aware in which provider data center the services reside, nor does the customer have to be aware. The beauty of these systems is that workloads can move around as necessary to maintain service level agreements.

Cloud service providers generally build their systems with the assumption that hardware will likely fail, which means that you, as the customer, can avoid the need to buy expensive failover and availability systems on your own.

For scale, the cloud provider can provide grid-like scalability to great levels so that you don't need to worry about how to grow when the time comes.

For public cloud, there are a number of pros and cons to consider. On the plus side, cloud will:

- Enable immediate implementation.

- Carry low to no initial deployment costs.

- Provide a consumption-based utility cost model.

- Provide more cost effective scale than would be feasible in a private data center.

However, there are definitely some downsides to cloud as well, which include:

- Potentially unpredictable ongoing usage charges

- Concerns around data location; many do not want data stored in US-based data centers due to concerns around the NSA and PATRIOT Act

- Charges across every aspect of the environment, from data storage to data transfer and more

- No control over underlying infrastructure

- Care needs to be taken to avoid lock-in

The faces of the public cloud

Here is a brief look at the different kinds of public cloud services that are available on the market.

Software-as-a-Service (SaaS)

From a customer perspective, software-as-a-service (SaaS) is the simplest kind of cloud service to consume as it is basically an application all wrapped up and ready to go. Common SaaS applications include Salesforce and Office 365.

With SaaS applications, the provider controls everything and provides to the customer an application layer interface that only controls very specific configuration items. Because all of the infrastructure and the fact that most of the software is hidden from the you as the customer, you don't need to worry about any underlying services except those which may extend the service, such as integrating Office 365 with your on-premises Active Directory environment.

Platform-as-a-Service (PaaS)

Sometimes, you don't need or want a complete application. In many cases, you just need a place to install your own applications but you don't want to have to worry at all about the underlying infrastructure or virtualization layers. This is where platform-as-a-service (PaaS) comes into play.

PaaS provides you with infrastructure and an application development platform that gives you the ability to automate and deploy applications including your own databases, tools, and services. As a customer, you simply manage the application and data layers.

Infrastructure-as-a-Service (IaaS)

In other cases, you need a bit more control, but you still may not want to have to directly manage the virtualization, storage, and networking layers. However, you need the ability to deploy your own operating systems inside vendor-provided virtual machines. Plus, you want to have the ability to manage operating systems, security, databases, and applications.

For some, infrastructure-as-a-service (IaaS) makes the most sense since the provider offers the network, storage, compute resources, and virtualization technology while you manage everything else.

On-Premises Reality

Even though public cloud has a number of desirable traits, there are some harsh realities with which CIOs and IT pros need to contend:

- **Security** – For some, particularly those in highly-regulated or highly-secure environments, the idea of moving to a multi-tenant public cloud is simply not feasible.

- **Bandwidth** – Many areas of the world remain underserved when it comes to bandwidth, and companies can't get sufficient bandwidth with sufficiently low latency to make cloud a feasible option.

- **Cost** – There may come a point at which cloud may become more expensive than simply building your own environment.

These challenges are reasons that many organizations are turning to private cloud environments.

Private Clouds

The term *private cloud* is often, well, clouded in confusion as people try to apply the term to a broad swath of data center architectures. So, let's try to clear up some of the confusion.

First and foremost, a private cloud environment generally resides in a single-tenant environment that is built out in an on-premises data center, but it can sometimes consist of a single tenant environment in a public data center. For the purposes of this chapter, we'll focus on the on-premises use case.

Private cloud environments are characterized by heavy virtualization that fully abstracts the applications from underlying hardware components. Virtualization is absolutely key to these kinds of environments. Some companies go so far as to offer internal service level agreements to internal clients in a cloud-like manner. The key phrase there is "internal clients" — that is the customer in a private cloud environment. For such environments, being able to provide service level guarantees may mean that multiple geographically dispersed data centers need to be built in order to replicate this feature of public cloud providers.

Heavy use of virtualization coupled with comprehensive automation tools reveals an additional benefit of private cloud: self-service. Moving to more of a self-service model has two primary benefits:

- Users get their needs serviced faster

- IT is forced to build or deploy automation tools to enable self-service functionality, thereby streamlining the administrative experience

As mentioned before, many companies want to keep their data center assets close at hand and in their full control, but they want to be able to gain some cloud-like attributes, hence the overall interest in private cloud. As is the case with public cloud, there are a number of pros and cons that need to be considered when building a private cloud.

In the pros column, private cloud:

- Provides an opportunity to shift workloads between servers to best manage spikes in utilization in a more automated fashion.

- Enables ability to deploy new workloads on a common infrastructure. Again, this comes courtesy of the virtualization layer.

- Provides full control of the entire environment, from hardware to storage to software in a way that enables operational efficiency. In other words, routine tasks are automated and repeatable.

- Allows customers to customize the environment since they own everything.

- Provides additional levels of security and compliance due to the single tenant nature of the infrastructure. Private cloud-type environments are often the default due to security concerns.

As with everything, not all is a perfect picture. Private clouds do have a number of drawbacks, including:

- Requiring customers to build, buy, and manage hardware. This is often something that many companies want to reduce or eliminate.

- Not always resulting in operational efficiency gains.

- Not really providing what is considered a cloud computing economic model. You still have to buy and maintain everything.

- Potentially carrying very high acquisition costs.

In short, private clouds are intended to have some of the architectural characteristics of public clouds while offering internal clients cloud-like economic outcomes when chargeback processes are implemented. Even if the central IT department providing the service doesn't really use "the cloud," as internal clients are able to provision and consume resources on demand — at least to a reasonable point — there is the beginning of a private cloud.

Hybrid Cloud

Increasingly, people are choosing both cloud options – public and private – to meet their needs. In a hybrid cloud scenario, the company builds its own on-premises private cloud infrastructure to meet local applications needs and also leverages public cloud where reasonable and possible. In this way, the company gets to pick and choose which services run where and can also move between them at will.

The Intersection of Cloud and Hyperconverged Infrastructure

If you're wondering what all of this talk about cloud has to do with hyperconverged infrastructure, wonder no more. Depending on the hyperconverged infrastructure solution you're considering, there are varying degrees of association between the hyperconverged infrastructure product and both public and private clouds.

Economics

Everything you've read so far leads to money. The potential to completely transform the data center funding model is one of the key outcomes when you consider hyperconverged infrastructure. With easier administration comes lower staffing costs. With the use of commodity hardware comes lower acquisition costs. With the ability to scale linearly in bite-size chunks, companies can get the beginnings of a consumption-based data center equipment acquisition model that enables closer to pay-as-you-go growth than traditional data center architectural models. As your environment needs to grow and as users demand new services, you can easily grow by adding new hyperconverged systems.

Scale

Agility implies some level of predictability in how workloads will function. Public cloud provides this capability. For those wishing to deploy a private cloud environment, these needs can be met by leveraging hyperconvergence's inherent ability to scale linearly (meaning, by scaling all resources including compute, storage, and networking simultaneously). In this way, you avoid potential resource constraint issues that can come from trying to manually adjust individual resources and you begin to achieve some of the economic benefits that have made public cloud a desirable option.

Scaling the data center should not result in scaling the complexity. In order to attain the full breadth of economic benefits that go with cloud, you have to make sure that the environment is very easy to manage or, at the very least, that management is efficient. This means that you need to automate what can be automated and try to reduce the number of consoles that it takes to get things done.

With hyperconverged infrastructure, management efficiency – even at scale – is a core feature of the solution. You are able to manage all of the elements included in the product from a single console and you are also able to apply a breadth of consolidated policies to virtual machines.

Geographic Diversity and Disaster Recovery

Also on the economics front, the value of disaster recovery cannot be overstated. One of the benefits of the cloud is the geographic diversity that can be achieved to protect against natural disasters. With a hyperconverged infrastructure solution that has data replication as a part of the core offering, multisite redundancy capability is baked in as part of the solution.

For those that have opted to build hybrid clouds, some hyperconverged infrastructure solutions can leverage that public cloud deployment as a replication target. In other words, rather than going to the expense of building out a second physical site, the public cloud can be used to achieve data protection goals.

Hyperconvergence and the Private Cloud

Building a traditional private cloud is hard. It takes a lot of work to get all the pieces aligned. Hyperconverged infrastructure can allow you to deploy private clouds in a fraction of the time it would normally take. Everything is built into the individual appliances, including centralized management, data efficiency, replication, and the ability to scale in incremental units. These are core needs in building an agile private cloud environment.

Up Next

You've now completed your introductory journey into the technical world of hyperconverged infrastructure. In the next section of this book, you will learn about a number of ways that you can begin use this knowledge in order to solve some of your most challenging business problems.

SECTION 2

Use Cases

6

Virtual Desktop Infrastructure

For years, IT pros have been trying their best to improve what has become a chaotic desktop management environment and to reduce costs for providing desktop computers. One of the original use cases around hyperconverged infrastructure was virtual desktop infrastructure (VDI).

VDI is an interesting solution. Like many trends in IT, VDI has gone through Gartner's "hype cycle" (**Figure 6-1**). It went through both a period of sky-high expectations and also hit rock bottom as people became increasingly disillusioned with the technology. Today, however, it's reaching the top end of the Slope of Enlightenment and entering the Plateau of Productivity.

How did we get to where we are?

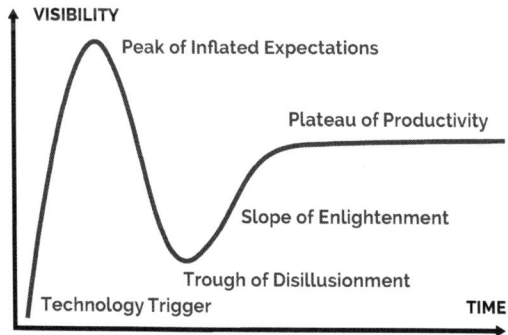

Figure 6-1: Gartner's hype cycle (courtesy of Wikipedia)

VDI Through the Years

Long before x86-based virtualization became the norm, IT departments searched for ways to simplify and streamline desktop computing. Microsoft and Citrix led the way in this space and, for a time, their products were ubiquitous. People deployed thin clients based on specialized editions of Windows Server and had an adequate experience. Unfortunately, their experience was one that was mostly useful where terminals, not full desktop capabilities, were needed.

Then came along server virtualization. Server virtualization resulted in the ability to transform the business and IT — lowering costs while increasing productivity and efficiency along the way. With server virtualization, data center administrators could almost completely replicate their physical servers inside software with little to no loss of functionality.

At some point, someone somewhere had the bright idea to attempt to apply the same thinking to desktops in order to close the user-experience gap and make terminal-based desktops more like their PC brethren. Things didn't work out quite so well. IT pros quickly discovered that their path to VDI success would be littered with very different challenges than those faced on the road to server virtualization.

VDI Workload Differentiators

Although servers and desktops are both computers, how they're used is very different. These differences have driven many of the challenges that doomed early VDI projects. Just because virtual desktops look like virtual servers, it doesn't mean they act like them. Whereas server-based workloads will have their own performance peaks and valleys, they're nothing compared to what happens in the world of the virtual desktop.

Types of Virtual Desktops

There are two different kinds of virtual desktops that you can use in a VDI environment: Persistent and Non-Persistent.

Persistent Desktops

Persistent desktops are the type that closely resemble desktop computers in the physical world. There is a 1:1 relationship between a virtual desktop and a user. In other words, a user has his or her own virtual desktop that no one else uses. This model is the most seamless from a user perspective since users have become used to having their own space. Persistent desktops require you to have sufficient storage for desktop customizations.

Non-Persistent Desktops

Think of a college computer lab: rows of computers available for any student, and students can go to different computers every day. The students really don't care which computer they use each day as long as one is available and they don't need to maintain user-specific settings. This is known as a non-persistent desktop. User settings are not maintained between sessions. Each time a user logs in, it's as if he or she has logged in for the first time.

Linear Usage Patterns

In VDI environments, usage patterns directly follow user actions. When users log in or boot their virtual desktops in the morning, each virtual desktop undergoes significant storage I/O operations.

Contrast this to a traditional PC, where you've probably seen it take minutes for computers to fully boot and login. This is because a lot of information has to be read from disk and placed into memory on a traditional PC. There are also write operations taking place, such as when Windows logs any exceptions that may take place at boot time.

Now, multiply all of this I/O by the number of users logging into their virtual desktops at the same time. In the world of the traditional desktop, each user has his or her own storage device (the local hard drive) to handle these I/O operations. In a VDI environment, the virtual desktops all share common storage systems, often a SAN or NAS device shared among the various hosts that house the virtual desktops. The amount of I/O that starts to hit storage can be in the hundreds, thousands, or even tens of thousands of IOPS.

The Failure and Resurgence of Storage

This was the problem in the early days of VDI. Then-current disk-based storage systems simply could not keep up with demands and quickly succumbed under the IOPS-based assault that came their way. This led directly to the hype cycle's Trough of Disillusionment as people quickly discovered that there would be no return on their VDI investment because they had to buy shelves and shelves of disks to keep up with I/O demands. In technical terms, getting appropriate performance characteristics wasn't cheap at all.

Shortly thereafter, flash storage started on its road into the enterprise. With the ability to eat IOPS faster than anything previously on the market, flash became a go-to technology for virtual desktops. It was used in many different ways, but flash carried its own baggage on the VDI journey. First, some of the flash-based solutions added complexity to storage, and second, all flash systems tended to be expensive.

Second-Class Citizenship for Data Protection

Protecting VDI environments was also a challenge. The nature of VDI didn't always mean that it would enjoy the same kinds of data protection services as server workloads, even though desktop computing really is a critical service. Between WAN bandwidth and backup storage needs, fully protecting the desktop environment wasn't always feasible.

It's All About That Scale

Scaling VDI was, again, a far different chore than scaling server-centric workloads. Whereas server workloads were scaled based on individual resource need, VDI-based workloads scaled far more linearly, requiring RAM, compute, and storage to scale simultaneously.

The User Experience Trumps All

Finally, let's talk about the user experience. In a perfect VDI world, you have persistent virtual desktops in which users' settings and experience are maintained between sessions. This is the scenario that most closely mimics the real desktop experience, and people like it. With legacy infrastructure, getting the performance and capacity needed to support persistent desktops can be a real challenge.

Many gave up on VDI, thinking that they would never be able to enjoy their dreams of an efficient desktop environment. But then something interesting happened. Hyperconverged infrastructure hit the market.

Hyperconvergence and VDI Scaling and Performance

As mentioned earlier in this chapter, VDI became one of the original primary use cases for the introduction of hyperconverged infrastructure into a company. It's not hard to see where hyperconvergence solved just about all of the challenges — real and perceived — around VDI.

First, let's talk about the ability for hyperconverged infrastructure to scale. You learned earlier that hyperconvergence natively enables linear resource scalability, which is also necessary for VDI environments to be able to keep pace with growth. As you add virtual desktops, you need to assign both CPU cores and RAM to those systems along with sufficient storage for the operating system, applications, and user files.

Performance is one of the big challenges in VDI, particularly as it relates to storage. With most hyperconverged infrastructure systems, you're getting a combination of flash storage and spinning disk. The flash layer is used to make everything faster while the spinning disk allows you to store user files on media designed for capacity. You get the best of both worlds with hyperconverged storage systems based on hybrid storage.

Further, with hyperconverged systems that have deduplication and compression features at the storage layer, you get even more benefits. Virtual desktops are all very similar, so they are very easily reduced at the storage layer. With reduction, you're able to store more virtual machines on the storage that exists in your hyperconverged infrastructure, which saves you a lot on disk costs. Deduplication and compression is the key technology that enables the use of persistent desktops in a VDI environment. Deduplication also massively reduced the I/O footprint for VDI systems. Being able to efficiently cache deduplicated desktop systems can virtually eliminate the various storms – boot storms and login storms – that can negatively impact performance otherwise.

Cache Is More Efficient When Deduplicated

By Brian Knudtson

In hyperconverged systems with full inline deduplication across all workloads, the data management layer tracks all of the individual references to the unique blocks that have been written to the hard disk drives (HDDs) in metadata.

This deduplication extends to data that is stored in cache. When placed into cache, a block is read from the hard disk drive (HDD), incurring a Read IO, and a copy is placed on the SSD drives. Although the block could be retrieved for a single request from a single virtual machine (VM), it could be requested again by the same VM (e.g., a block used by multiple files) or even requested by a different VM on the same host (e.g., a core Windows file). When the second request comes looking for the same block, it has already been placed in cache. Now you have two VMs benefiting from a single cached block. Extend this example to 10 VMs, and we have a single block in cache that could be worth 10 or more blocks in a non-deduplicated environment.

Continuing this example, the nine additional blocks that were accessed directly from cache generated no I/O to the back-end disks. That's nine less HDD IOPS consumed that are now available for another read or write operation that hasn't been cached already.

Imagine the benefits that a VDI environment could realize during a boot storm. All the VMs are based on the same template, and therefore they all have the same set of files during initial boot. Normally, 100 VMs all booting at the same time would require a significant number of HDDs, but with this hyperconverged infrastructure platform, the first VM to boot reads the block off the HDD, which promotes that block into cache. Now the next 99 VMs can all access that same block from cache. That's a 100:1 IOPS reduction on the IOPS-bound disks.

Data Blocks

Inline Deduplication Processing

On Disk

Deduplication in cache

Some hyperconverged systems have caching solutions that utilize RAM as a tier. In such cases, you can now apply all the same benefits just discussed and use a speed of memory-based cache. Blocks used by 100 VMs brought into memory cache at the cost of only a single VM can bring a significant performance increase to a VDI environment.

Another huge advantage for cache that comes from tracking all data as meta-data is more intelligent cache-warming algorithms. Instead of simply grabbing the next block on the disk and betting on previous writes being sequential (see the IO Blender effect), the hyperconverged infrastructure nodes will calculate predictive caching based on the metadata. This leads to a much more intelligent and successful predictive cache approach.

This is just one of many advantages hyperconverged infrastructure can provide. This data efficiency, when applied to cache, not only improves the space utilization of cache, logically providing a larger cache, but also prevents read operations from going to the HDDs. It helps you improve your application performance and realize, on average, 40:1 data efficiency.

Let's not forget about data protection and availability. In a traditional desktop environment, fully protecting workstations can be a tough task and, in the event that a workstation happens to fail, a user could be without a computer for an extended period of time. In a VDI environment, if a user's endpoint fails, it can be very quickly replaced with another endpoint —the user simply reestablishes a connection to the persistent desktop.

But data protection in VDI goes way beyond just making it easy to get users back up and running. In fact, it comes down to being able to fully recover the desktop computing environment just like any other mission-critical enterprise application. In a hyperconverged infrastructure environment with comprehensive data protection capabilities, even VDI-based desktop systems enjoy backup and replication for users' persistent desktops. In other words, even if you suffer a complete loss of your primary data center, your users can pick right up where they left off thanks to the fact that their desktops were replicated to a secondary site. Everything will be there — their customizations, email, and all of their documents.

Up Next

With easy scalability, excellent performance capabilities, and great data protection features, hyperconverged infrastructure has become a natural choice for virtual desktop infrastructure environments. Up next, let's go a little deeper into how this architecture can help you address remote office and branch office scenarios.

7

Remote Office/ Branch Office

Remote Office and Branch Office (ROBO) IT can create some pretty tough situations for IT pros to conquer. Perhaps the most significant problem is one of scale. Many ROBO environments have the need to grow very large, but need to do so by remaining very small. Consider a fictional company that has 500 locations. This is a big company, so the overall aggregate technology needs of the organization will be significant. At the same time, though, each branch location in this company supports 20 employees and is also a sales outlet. The individual sizing needs of each branch are relatively modest.

Traditional ROBO Challenges

At first look, it might seem like a simple solution. Throw a couple of servers into each location and call it a day. Unfortunately, it's not that easy. There's a lot more to the scenario than meets the eye.

First and foremost, just a couple of servers may not meet the needs of the branch office. Each branch is probably sized a little differently, so some may be able to operate with just a couple of servers while others may need more substantial storage capabilities. You'll probably want two or more servers just so you can have some level of availability. If one server fails, the other one can pick up the load. Getting high availability with just two servers, while solvable, isn't always straightforward.

At the same time, you have to keep an eye on performance to make sure that poorly performing local applications don't negatively impact the branch's business. You can't forget about data protection, either. If this was a single-site company, data protection would be relatively easy; you just back data up to a tape, disk, or a second location. But if you have many sites and some have slow Internet links, it can be tough to protect data in a way that makes sense. You don't want to have local IT staff that needs to change tapes, or watch backup appliances. You also don't want to have non-technical people trying to do this as a part of their jobs. It doesn't always work out well.

Plus, there's ongoing support. Stuff happens. You need to be able to keep every site operational. However, with each site you add (each with its own unique needs), the overall complexity level can become overwhelming. As complexity increases, efficiency decreases and it becomes more difficult to correct problems that might occur. Figure 7-1 provides a demonstrative overview of today's data center. In many ROBOs, centralized IT delivers services to the remote sites from a centralized location over a WAN. By centralizing IT, the company eliminates the cost of skilled IT staff on site at remote sites and reduces the risk to business continuity since IT handles data protection. However, the major drawbacks are often poor application performance, scattered management, and difficulty correcting issues that arise in remote sites.

To summarize the challenges faced in ROBO environments:

- There is a need for a lot of decentralized systems to support individual branch offices, and there is often lack of a cohesive management platform.

- Bandwidth to branch offices can often be limited and may not be reliable. Most ROBO sites lack the full breadth of data center services (high-performance storage, WAN accelerators, etc.) enjoyed by headquarters and by single-site companies.

- Data generated at branch offices needs first-class citizen protection, but often can't get it using legacy tools.

- Hardware at branch offices might run the gamut from just a server or two to a full cluster with a SAN, but most companies want to be able to have minimal hardware at branch locations when possible and need to be able to centrally manage solutions.

- There is a lack of technical personnel at most ROBO locations and companies don't want to have to hire dedicated technical staff for each one.

Figure 7-1: Chaos is the norm in many ROBO environments

Without some kind of change, the dystopian future for the ROBO will be so challenging that even Katniss Everdeen would call it quits and hang up her bow and arrow.

Transforming ROBO Operations with Hyperconverged Infrastructure

ROBO operations is an area in which the right hyperconverged infrastructure solution has the potential to completely transform the environment and how that environment is managed. The overall results can be lower costs, improved efficiency, and better overall disaster recovery capabilities.

So what does it take to achieve this ROBO utopia and how does hyperconverged infrastructure fit into the equation?

Keeping IT Simple

Hyperconverged infrastructure brings simplicity to chaotic IT organizations and nothing says "chaotic" like dozens of different sites running disparate hardware managed as individual entities. By moving to a common hyperconverged infrastructure platform, you instantly gain centralized administrative capabilities that encompass every site. Moreover, when it comes to hardware support, every site becomes a mirror of the others, thereby streamlining your support efforts. Such an architecture eliminates the need for dedicated technical staff at each branch.

The need to keep management simple cannot be overstated. Companies are no longer willing to scale IT staff at the same rate that they add sites and services, but they expect consistent performance. To solve this seeming paradox of intentions, IT has to look at the ROBO environment much more discerningly and deploy solutions. They need to choose solutions that unify management across all ROBO sites in a way that makes them appear as if they're a single entity even while they support a dispersed organization.

Less Hardware

Some sites need very little hardware while others need more. Some sites traditionally needed dedicated storage while others didn't. It's chaos. With the right hyperconverged infrastructure solution, you can have sites that operate on just one or two appliance-based nodes without having to compromise on storage capacity and performance. You simply deploy the two nodes, and they form their own cluster that imbues the branch with highly efficient storage capabilities that include comprehensive data deduplication and reduction. For larger sites, you simply add more nodes. No SAN is needed and all of the hardware across all of the sites is common, easy to support, and provides enterprise-level capabilities, even in a single-node or two-node cluster.

The data reduction features available in some hyperconverged infrastructure solutions mean that you don't need to constantly add storage. With reduction, you get to cram more data into the same amount of overall capacity at the branch site. Reduction also has other benefits. Read on.

Comprehensive Data Protection

Data generated or managed at branch sites needs to be treated just like data generated at HQ. In many cases, the data at branch sites is even more important because it's the information that's created as the result of sales or other customer-facing efforts. With a hyperconverged infrastructure solution that has the ability to fully compress and deduplicate data and that can work with data in its reduced form, you can get data protection capabilities that allow you to replicate branch office data to other branches or to headquarters even over slow WAN connections. Better yet, you don't need WAN accelerators to accomplish this feat. With the right solution, reduced data is transferred over the wire and, even then, only the blocks that don't already exist at the target site are transferred, resulting in an incredibly efficient process. This kind of data protection infrastructure also eliminates the need for on-site staff to perform tasks such as changing tapes and increases the potential for successful recovery in the event of a disaster. In **Figure 7-2**, you see a nice, neat, and streamlined infrastructure.

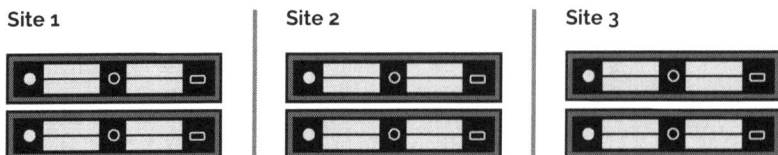

Site 1   **Site 2**   **Site 3**

Figure 7-2: Hyperconverged infrastructure tames the ROBO chaos

Deployment Options

As you're deploying ROBO solutions using hyperconverged infrastructure, you need to take a look at how you want your ROBO sites configured. There are two typical models available:

- **Hub and spoke**
  (**Figure 7-3**). With this architecture, there is a centralized hub in the center and each remote site is at the end of a spoke. With this model, the various remote sites will generally talk to the central hub, but not often with each other. Backups and other data transfer operations will generally flow from the end of one of the spokes back the hub.

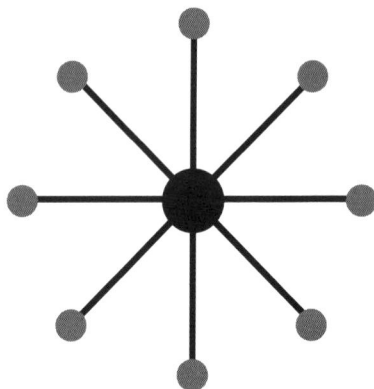

Figure 7-3: A look at a hub and spoke ROBO model

- **Mesh (Figure 7-4).** In a mesh environment, all of the sites can talk directly to the other sites in the mesh. Under this model, it's possible to have individual sites back up to each other and the organization can, theoretically, operate without a centralized hub, although one of the nodes often acts in this capacity.

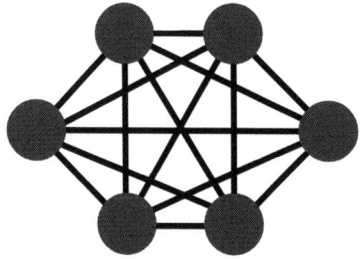

Figure 7-4: A look at a mesh-based ROBO model

As you're deploying hyperconvergence throughout your organization, it's important to ensure that the intended solution can easily support whichever deployment model you use, even if it happens to be a combination of the two. Most importantly, regardless of which model you use, you should be able to centrally manage everything and have the ability to implement data protection in whatever way makes the most sense for you. Finally, adding new sites – scaling the environment – should be a basic feature of the solution and not a complicated afterthought.

Up Next

ROBO environments can be considered an "application" that requires some specialized attention. This is a perfect use case for hyperconvergence. However, there are some actual applications that have special resource needs as well. In the next chapter, you'll learn about these applications and how hyperconvergence helps.

8

Tier 1/Dedicated Application Support

Not every company needs to tear down their entire data center and replace everything with shiny new hyperconverged infrastructure appliances. The chances are pretty good that you can't really do that even if you wanted to. However, you may have a single application that's challenging you and needs to be tamed. Or, perhaps you have a new application that you need to deploy, and you can't deploy it on your existing data center infrastructure.

For you, hyperconverged infrastructure still might be just the answer. In fact, even if you have only a single application, you might still be able to leverage hyperconvergence.

Enterprise Application Needs & Challenges

Not all enterprise applications are created equal. Every application has a unique performance profile, and each requires a varying amount of resources to be dedicated to that application. In Chapter 6, you learned about the popular enterprise application

Virtual Desktop Infrastructure (VDI), and discovered that it has very different resource needs than general server virtualization.

Most traditional data centers are not equipped to handle applications that don't fit a mainstream operational envelope. That is, most traditional data centers are equipped to operate a broad swath of mainstream applications, but don't always have the capability to support applications with very unique resource needs. The kinds of applications that fit into this category will vary dramatically from company to company. For some, the entire centralized IT function consists of just a file server, so even something as common as an Exchange system would place undue stress on the traditional environment. For others, the traditional environment handily supports Exchange, but SQL Server is a step too far.

Every application has some kind of an I/O profile. This I/O profile dictates how the application will perform in various situations and under what kind of load. On top of that, every organization uses their systems a bit differently, so I/O profiles won't always match between organizations. As you deploy new applications, it might be time to leverage hyperconverged infrastructure.

A lot of people worry about virtualizing some of their resource hungry applications for fear that they won't perform well. This is why, even to this day, many companies still deploy physical SQL Server, Exchange, and SharePoint clusters. While physical deployment isn't "wrong," the benefits of virtualization are well-known and include better overall hardware utilization and better data protection capabilities.

Hyperconvergence and Dedicated Applications

The right hyperconverged infrastructure solution can help you to virtualize even the largest of your Tier 1 mission-critical applications while also ensuring that you have sufficient resources to operate these workloads. Plus, don't forget the major role hardware acceleration plays in some hyperconverged systems.

By offloading the "heavy lifting" operations, you can more confidently virtualize I/O-heavy applications while also reducing the amount of storage capacity those applications require. With deduplication being handled by a hardware card, you can gain the benefits of deduplication without incurring the typical performance penalty that can be introduced when deduplication has to be handled by a commodity processor.

Elements of the Microsoft stack, including SQL Server and SharePoint, can be safely virtualized and significantly accelerated by moving to hyperconvergence. The same holds true for Oracle. Other I/O hungry applications are growing in popularity, too. Two emerging applications that carry with them pretty significant I/O requirements are Splunk and Hadoop. Splunk is a logging tool subjected to abusive write needs while Hadoop is a big data analytics tool that requires a whole lot of both read and write I/O capability. Both need a lot of storage capacity, too, which is where the aforementioned deduplication features come into play.

Even better, as you need to grow, you just grow. Scalability is a core part of the infrastructure. When you grow, you can add more storage capacity, more storage performance, more CPU, and more RAM as needed, so you don't need to worry about encountering a resource constraint somewhere along the line. That said, one common misperception about hyperconverged infrastructure is that you are absolutely required to scale all resources at exactly the same rate. This is simply not true. For example, with SimpliVity, you can add compute-only nodes that don't have any storage. With Nutanix, you can add storage-heavy nodes. It's not a one-size-fits-all conversation. Even in these cases, scaling is simple.

Moreover, for whichever applications you choose to include in your hyperconverged infrastructure, depending on the hyperconverged infrastructure solution you select, you can gain comprehensive data protection capabilities that will help you more quickly recover in the event of a disaster or another incident. In addition, you can also inherit the ability to manage the hyperconverged environment from a single administrative console.

Finally, if you're thinking "private cloud" with regard to your data center, you *have* to virtualize your Tier 1 applications in order to bring them into the centralized, API-driven management fold. A private cloud is a VM-centric construct that requires high levels of virtualization to imbue the environment with the agility and flexibility needed to get things done.

Up Next

Just when you thought that you had everything solved by virtualizing and moving to hyperconvergence all of your Tier 1 applications, now comes along a directive to consolidate your disparate data centers. That, dear reader, is the topic of Chapter 9.

9

Data Center Consolidation

Mergers and acquisitions. Cost cutting. New business initiatives. There are all kinds of reasons why companies make the decision to consolidate data centers. Maybe your company undergoes explosive, barely controlled growth or they may decide to stop, take a pause, and reconsider how IT does business in their organization. Or maybe your company decides to buy out another company, and you suddenly inherit a whole series of data centers that you're not prepared to handle.

Want to know a secret? It will be up to you to figure it out. Furthermore, you'll probably be asked to do it with the same budget you already have.

Here's the thing, though. Data center consolidation isn't always just about reducing the number of data centers from a big number to a smaller number. Sometimes, it's about reducing the amount of stuff strewn about the data centers that you already have.

Today's IT organizations generally buy and integrate numerous point solutions, from a plethora of vendors, each with its own training courses, licensing, refresh cycles, and modes of operation.

These point products are the result of years of planning and investments to support business applications.

We hear the same story time and time again. Does this sound familiar? You virtualized about 5 to 8 years ago and naturally a data protection strategy project came directly after that. The Share-Point implementation project for your marketing organization took eons to complete and required purchasing a new SAN. It feels like you just bought that SAN yesterday, but you blinked, three years flew by, and it's time to refresh . . . again. The decision to buy all of these products made sense at the time, but today data center complexity can feel overwhelming and discourage innovation.

All of this has really cooked up several challenges for IT organizations, including:

- Time overhead spent on operational tasks

- Mobility and management of virtual machines (VMs)

- Budget constraints

- Breaching service level agreements

- Operational efficiency

- Application performance

The list goes on and on.

Ask yourself these questions to determine where you might have pain points:

- As data growth explodes, can you continue to operate the same way you always have?

- Are your legacy technologies designed for virtualized environments?

- How much time are you spending on maintenance, upgrades, deployments, provisioning and management tasks instead of building more valuable innovation for the business?

- Do you have the necessary expertise to manage each of these products separately?

- Are you struggling to meet your Service Level Agreements (SLAs) with the business?

- Are you missing data protection objectives like Recovery Point Objectives (RPOs) and Recovery Time Objectives (RTOs)?

If you answered "No" to one or more of these questions, you might be a candidate for hyperconverged infrastructure.

Consolidation with Hyperconvergence

In every chapter of this book so far, you've learned about how hyperconverged infrastructure solutions can reduce the variety of hardware and software you have to manage in the data center. Every time you eliminate a class of hardware or software in your data centers, you are on your way to answering "Yes" to all of the questions outlined in the previous section.

That's the ideal scenario.

At the most basic level, hyperconverged infrastructure consolidates storage and compute, enabling you to eliminate the monolithic SAN environment. From there, some hyperconverged infrastructure vendors make things pretty interesting. For example, Nutanix and SimpliVity both provide something in the way of data reduction via deduplication and compression. SimpliVity, however, takes this to the extreme through the use of their accelerator card, which forms the basis for what they call their OmniStack Data Virtualization Platform.

By enabling global inline deduplication and compression with a solution like SimpliVity, you suddenly need less overall capacity, which means you need less overall hardware. With constant data reduction, you no longer need:

- WAN accelerators to reduce data over the wide area network because data stays reduced

- Separate backup software to protect the data in your environment

- Separate deduplication appliances

- SSD arrays

Instead, you can massively reduce the amount of hardware and software that you're operating, maintaining, getting trained on, and, maybe even struggling with. With less stuff to manage and worry about, you can better focus on the business and on improving SLAs, RTOs, and RPOs. You get to focus on the business rather than on the technology.

If you're in a situation in which you need to reduce the number of data centers you're managing, hyperconverged infrastructure can help you there, too. How? For the same reasons that we just discussed. Rather than just taking all of the hardware from the various sites and combining it all into one supersite, you can rethink the whole model. In addition to cutting down physical locations, you can also minimize complexity.

Up Next

Data center consolidation is important, but you still need a place to run applications and perform testing and development. Testing and development environments are often short-changed in the world of IT. In the next chapter, you will learn why that's *not* a good situation. Plus, you'll learn how hyperconverged infrastructure can help you to improve your operations from testing to production.

10

Test & Development Environments

It's pretty clear that production environments enjoy premier status in most data centers. Production gets the fastest storage, the biggest servers, and all of the supporting services which make the application magic happen. Meanwhile, the poor test and development (test/dev) environment doesn't get all that much attention.

Let's take a look at what the test/dev environment supports. Test/dev consists of important activities, which include:

- Testing new application versions as they are released in order to determine potential impact on production.

- Creating new custom software to serve the needs of the business.

- Having a place to perform unit testing and load testing for new software being created by developers.

In fact, in some organizations, even the developers' development machines are virtualized, and they work against virtualized instances of production software to ensure that their efforts will translate well into production.

The State of Test/Dev Environments

In many companies, test/dev environments are often given leftovers and hand-me-downs. For example, production servers that have been decommissioned might be moved to the test lab or to a development lab. These servers are configured just like they were three to five years ago when they were originally purchased, and they generally do not have warranty support. Further, they use hardware that is one or two generations removed from current products.

The same goes for your storage systems that support the test/dev environment. Storage might consist of the old SAN that was removed from production. Or it might include a cheap array of disks which provide reasonable capacity but is lacking performance.

At first, this may seem like a reasonable thing to do. After all, test/dev is a lower priority than production, right? Well, there are a few reasons why test/dev is more important than you might think:

- **Time is money** — That's the old adage. By using older, slower equipment in test/dev, you waste staff time that could be better spent doing other things.

- **Development efficiency** — Your developers are likely among your higher paid staff. The more you short-change their work environment, the slower they work and the less efficient they become. This leads to slower overall development time and increases time to market for new features, products, and services.

- **Work stoppage** — Not having a warranty equates to having non-existent or slow service, if and when equipment fails. Failure in test/dev means that a critical piece of your environment is no longer available.

The Impact on Production

In most organizations, it's good to make sure that the test/dev environment resembles the production environment, especially when it comes to developing software and pushing it from test/dev to production. When there is massive variance between test/dev or when test/dev is not sufficient, bad things can happen, including:

- **Perplexing performance** — An inability to truly determine how well an application will perform in production means that you can't quickly resolve performance related problems. When hardware between production and test/dev isn't close, applications will probably run very differently. This means you can't easily predict how well applications will operate.

- **Elevated expenses** — Some say that having underpowered hardware in test/dev actually makes sense since it means that, if an application performs well there, it's guaranteed to work well in the more robust production environment. In essence, they're saying that overbuilding production makes sense. That means that you're buying resources you may not need.

- **Insidious inefficiency** — The fact is that having two complete sets of hardware doesn't always make sense, even when it's necessary.

- **Dubious data defense** — Many people don't do data protection in test/dev since it's not as critical as production. For those that do a lot of internal development, they often do take steps to protect code, but not always to the level that they do in production and they may leave test/dev more vulnerable than they would like.

When There Is No Test/Dev

There are companies that don't have *any* test/dev environment. They don't have the budget, the personnel, or the space to stand up a complete test environment, so they operate by directly updating production before performing complete testing. This is a relatively high risk activity that can be disastrous if a mistake is made. We recommend having at least *some* kind of testing capability to make sure that updates to production don't result in downtime.

Hyperconverged Infrastructure in Test/Dev Environments

Once again, the right hyperconverged infrastructure solution has the potential to address all of the challenges identified in the previous section. Further, with the right solution, you can also add test/dev capability to companies that may not have had it in the past.

There are a couple of ways you can stand up a test/dev environment using a hyperconverged infrastructure solution:

- Build a separate environment

- Add an additional node or two to production

- Use hyperconverged infrastructure for test/dev only

Each of these methods has its own benefits. Building a complete environment that mirrors production makes it possible to truly see how well applications will perform in the production environment and also provides plenty of capacity to allow development to take place. **Figure 10-1** gives you a look at how such an environment might be structured.

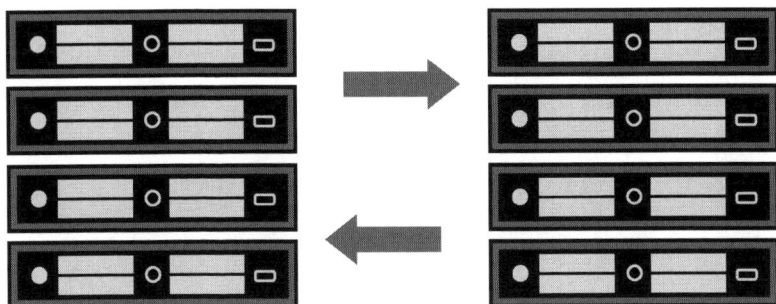

Figure 10-1: Building out two environments to support separate test/dev and production scenarios

Further, this method makes it possible to use each hyperconverged environment as a replication target for the other. You can protect production by replicating it to test/dev and vice versa. When you stand up the environment like this, you can also take advantage of any global deduplication capabilities offered by your hyperconverged infrastructure platform.

This is a key factor in containing costs. In essence, you can deduplicate the *entire* environment, and, since test/dev mimics production, the capacity savings can be huge.

If you don't need a complete replica of production, you can also opt to simply add an additional node to your existing production hyperconverged infrastructure environment. As is the case with building a completely separate test/dev environment, you will still enjoy the incredible capacity savings that come with global deduplication. This benefit is also for the same reason —the test/dev workloads mimic production, so even though there are a lot of identical blocks floating around the workloads, each of those only has to be stored one time.

When it comes to disaster recovery, you have a few options as well. With a separate environment scenario, you already know that you can do disaster recovery between the two environments via replication. With a hyperconverged infrastructure solution that deduplicates across the entire environment, you will save a whole lot of storage capacity. Also, if you choose to simply add nodes to production to handle test/dev needs, you can still replicate everything from production to a secondary site if you have one. Regardless, you will still be able to withstand the loss of a node in the cluster while maintaining operational production and test/dev capabilities.

There is also a third potential use case: using hyperconverged infrastructure for test/dev only. It's entirely possible that you already have a well-running production machine and you don't want to move it to hyperconverged infrastructure. There's not a reason that you can't consider using the architecture for test/dev only. This will ease the administrative burden in test/dev and avoid the need to get too deep into the technical weeds for that environment. Things will just work. You won't need to buy a separate SAN and you'll get very good performance for this critical infrastructure arena. Further, since you will probably have a lot of different copies of the similar virtual machines in test/dev, you'll be able to get great benefit from any data reduction services that may exist in the hyperconverged infrastructure solution.

With hyperconverged infrastructure, you will not have to maintain the IT skills around the dev/test SAN and other needs and will be able to focus the development budget on application development. There are also some other benefits that can be had by using hyperconverged infrastructure in test/dev:

- Allows you to keep pace with business needs by quickly turning around incremental tasks in a production-like environment

- Ability to clone production environments and integration environments in minutes

- Having a well defined process that includes ways to push changes to production including creating backup of the original environment to have ability to roll-back

- Developers and businesses would like to adopt a SaaS model for test/dev and are looking for cloud like elasticity and ease of getting environments established

Up Next

It's clear that test and development environments can be significant assets, but they're only useful if they're leveraged in a way that supports the needs of the business. In the next chapter, we'll talk about what happens when IT goes rogue . . . or at least appears to. Alignment between IT and the business is the topic of Chapter 11.

11

Aligning Architecture & Priorities

Perhaps one of the most enduring meta-conversations about IT in past decades has been around how well IT serves the needs of the business. Often referred to as "IT/business alignment" this conversation generally used to indicate when IT failed to meet the needs of the business. In a perfect world, there wouldn't have to be this conversation because IT would never be considered as off-track or "rogue." Unfortunately, that is not reality. IT often struggles to maintain a focus on the business, a problem often exacerbated by the infrastructure solutions that have been adopted.

In fact, this whole idea of alignment is one that hyperconverged infrastructure has the potential to address head-on. No, it won't fix every alignment problem in every organization, but it can begin the process.

The 2015 State of Hyperconverged Infrastructure

In 2015, we at ActualTech Media published a report entitled *The 2015 State of Hyperconverged Infrastructure*. In researching this report, we uncovered a pretty significant misalignment between IT priorities and the potential hyperconverged infrastructure benefits.

Download your copy of the full report

If you'd like to get an in-depth look at the results of this survey, the full report is available for free download at www.hyperconverged.org.

Data centers are among the costliest physical assets owned and operated by organizations. The cost is not just in the equipment that is deployed, but also in the sheer effort that it takes to manage that equipment, keep it running, and keep it maintained year after year. To make matters worse, many companies have deployed Band-Aid-like solutions to patch over problems introduced as the data center grows more complex or is challenged to meet emerging business needs.

Let's start with the items considered priorities by respondents. In **Figure 11-1**, you will see that improving data protection, improving operational efficiency, and implementing VDI are the top three items on respondents' radars. Remember, these responses do not consider the role of hyperconverged infrastructure; these are simply overall IT priorities.

Now, let's look at respondents' primary driver for considering hyperconverged infrastructure, the results of which are shown in **Figure 11-2**. See if you can tell exactly where the results of each question diverge from one another. Notice anything interesting?

Which of the following would you consider to be your organization's most important IT priorities over the next 12 to 18 months? (Multiple Responses Allowed)

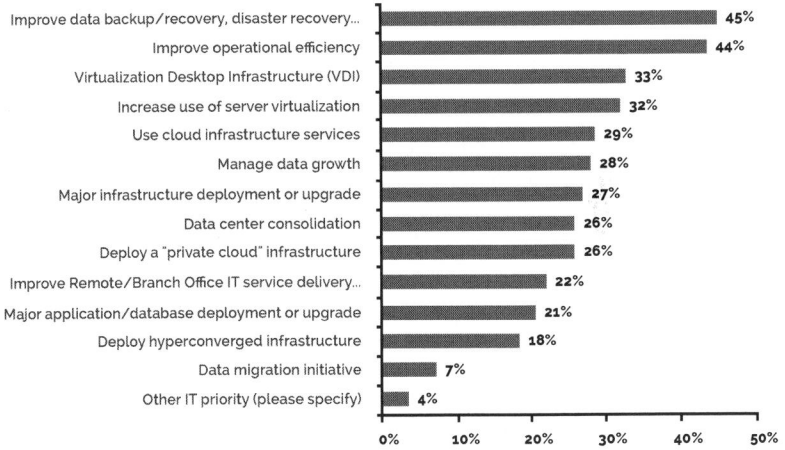

Priority	%
Improve data backup/recovery, disaster recovery...	45%
Improve operational efficiency	44%
Virtualization Desktop Infrastructure (VDI)	33%
Increase use of server virtualization	32%
Use cloud infrastructure services	29%
Manage data growth	28%
Major infrastructure deployment or upgrade	27%
Data center consolidation	26%
Deploy a "private cloud" infrastructure	26%
Improve Remote/Branch Office IT service delivery...	22%
Major application/database deployment or upgrade	21%
Deploy hyperconverged infrastructure	18%
Data migration initiative	7%
Other IT priority (please specify)	4%

Figure 11-1: Top IT Priorities

Which is the primary driver for your interest in hyperconverged infrastructure?

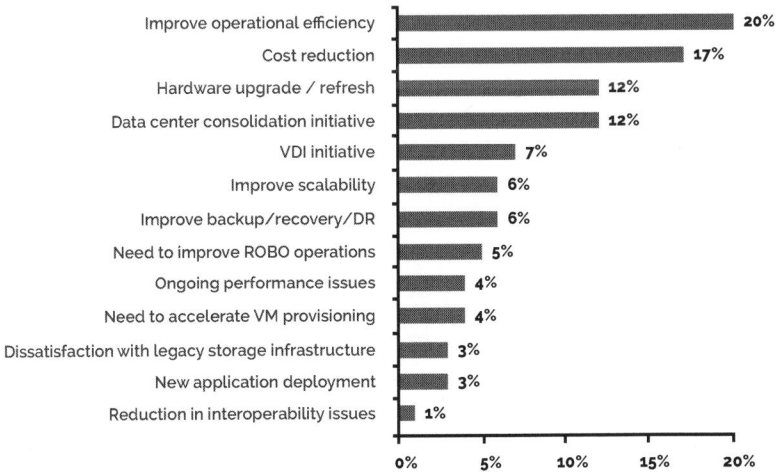

Driver	%
Improve operational efficiency	20%
Cost reduction	17%
Hardware upgrade / refresh	12%
Data center consolidation initiative	12%
VDI initiative	7%
Improve scalability	6%
Improve backup/recovery/DR	6%
Need to improve ROBO operations	5%
Ongoing performance issues	4%
Need to accelerate VM provisioning	4%
Dissatisfaction with legacy storage infrastructure	3%
New application deployment	3%
Reduction in interoperability issues	1%

Figure 11-2: Primary Driver For Interest in Hyperconverged Infrastructure

Improving operational efficiency is near the top of both lists, as we'll discuss later in this chapter. What's a bit more interesting is where we see divergence, particularly as it pertains to data protection. There is a vast gulf between the importance of data protection on the overall IT priorities list and what people look for in hyperconverged infrastructure.

Data Protection

Improving data backup and disaster recovery emerged as the single most important overall need for the IT organization from this research. In comparing key drivers for hyperconverged infrastructure against larger IT initiatives, it was surprising to see that data protection ranked seventh in the list despite the fact that it was identified as the highest IT priority to address. This may be due to the fact that enterprises are not equating modernizing the architecture with hyperconverged infrastructure with modernizing data protection; they continue to view hyperconverged solutions as simple conglomerations of servers and storage. Since, to many people, "hyperconverged" simply means exactly that, it may not be so far-fetched that they do not consider data protection a key part of the hyperconverged package. Many hyperconverged infrastructure solutions include backup, recovery, disaster recovery, and business continuity capabilities.

This book devotes an entire chapter to this topic, so we won't reiterate all of that here, except to say that those who have significant backup, recovery and disaster-recovery needs would do well to carefully study the hyperconverged infrastructure market and understand what's possible in this realm. With the right solution, there are some impressive data protection capabilities available.

Operational Efficiency

The virtual machine (VM) is the center of the universe when it comes to applications in most modern data centers. Most new workloads are deployed in VMs. However, consider the state of centralized policy in the data center. For data centers that have equipment from a wide variety of vendors, or that have a lot of "point solutions" (such as WAN accelerators and replication tools), there could be a number of touch points when it comes to policies.

These various touch points don't always align very well with one another, particularly when there are different vendors in the mix. For example, while it may be possible to define some policies at the hypervisor layer, it's often difficult to apply storage policies that have any awareness of VM boundaries. There are myriad other devices in the data center that can suffer from the same problem.

Since the VM is the center of the data center universe, why not implement a system that focuses directly on these constructs? Hyperconverged infrastructure solutions provide this opportunity to varying degrees, depending on vendor. Rather than go to three different places to define storage, backup, and replication policies, some hyperconverged infrastructure systems enable these policies to be attached to the VM.

Policy application is just one aspect of operational efficiency. There are many more, including:

- **Shielding complexity from the administrator** — Even IT pros shouldn't be subjected to complexity in the infrastructure when it can be avoided. Hyperconverged infrastructure helps make this happen. Availability mechanisms, such as RAID configurations and management, are often hidden from view and are simply a part of the environment.

- **Use-case improvements** — Ensuring that new applications and use cases, such as ROBO and VDI deployments can be supported without adding complexity and introducing inefficiency into operations, is critically important to help IT maintain alignment with business needs. When deploying these kinds of applications introduced inefficiency, IT and business alignment will suffer.

- **Overall alignment enhancement** — As has been mentioned, efficiency and simplicity can help IT better achieve alignment with the business.

You will have noticed that cost reduction is also very high on the list for survey respondents. We believe that cost reduction and operational efficiency go hand in hand with one another for many people. However, we also understand that hyperconvergence has the potential to dramatically improve how the IT budget is constructed. You'll learn much more about the economics behind hyperconvergence in Chapter 13.

Up Next

Alignment is about more than just technology. It's also about people and hyperconvergence can and will have some impact on you. Don't let hyperconvergence worry you with regard to what it means for your job. In the next chapter, we'll tackle that issue.

SECTION 3

Organizational Considerations

12

Overcoming Inertia — Mitigating Hyperconvergence's Perceived Challenges

---

In the *2015 State of Hyperconverged Infrastructure Market report*, IT pros identified a number of reasons why they are reluctant to move forward with hyperconverged infrastructure initiatives (**Figure 12-1**). Many of these reasons were exposed in the State of Hyperconverged Infrastructure report, but there are also other concerns that people have around hyperconvergence, particularly when it comes to potential job impact.

Not everyone is enamored with the potential for hyperconverged infrastructure. Some simply have no present need to re-examine data center operations. There are myriad reasons why respondents identifi ed these challenges around hyperconverged infrastructure adoption.

What is the primary reason that you have no interest in deploying hyperconverged infrastructure in the near term? (N=177)

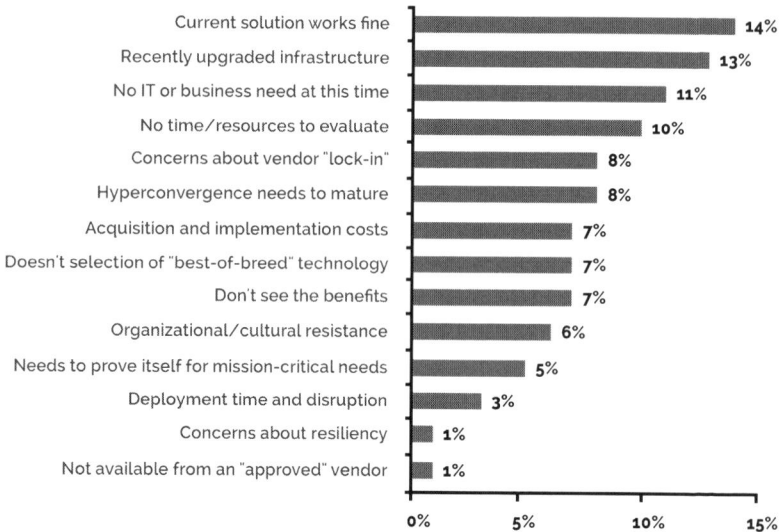

Reason	Percentage
Current solution works fine	14%
Recently upgraded infrastructure	13%
No IT or business need at this time	11%
No time/resources to evaluate	10%
Concerns about vendor "lock-in"	8%
Hyperconvergence needs to mature	8%
Acquisition and implementation costs	7%
Doesn't selection of "best-of-breed" technology	7%
Don't see the benefits	7%
Organizational/cultural resistance	6%
Needs to prove itself for mission-critical needs	5%
Deployment time and disruption	3%
Concerns about resiliency	1%
Not available from an "approved" vendor	1%

Figure 12-1: Primary Reasons for Not Deploying Hyperconverged Infrastructure

However, the top reasons why people aren't yet looking at the technology have nothing to do with the technology itself but rather have to do with the business cycle:

- **Current solution works just fine** — The adage "If it's not broke, why fix it?" holds true for many. However, that won't always be the case. Business priorities change on a dime, and understanding the significant benefits that may come from modernizing the data center with hyperconverged infrastructure solutions will be useful for the future.

- **Recently upgraded infrastructure** — For most hyperconverged infrastructure vendors, there is no need for a forklift. Different solutions offer varying degrees of integration opportunities but most can integrate into the existing environment at some level. Whether new applications are being deployed or there is a specific use case (such as VDI or

ROBO), there may be an opportunity to introduce hyper-converged infrastructure into the environment.

- **No current IT or business need** — Some people truly have no present infrastructure needs and are focusing their efforts in other areas. Again, this won't always be the case though, and these folks can sometimes be considered as part of the "current solution works just fine" category.

In this chapter, we're going to tackle some of the more serious issues that people have brought up with regard to hyperconvergence.

The IT Staffing Challenge

We hate to be bearers of bad news, but most companies want to limit the number of IT staff that they hire, at least in certain areas. Infrastructure is almost always one of those areas because it's directly associated with the expense side of the ledger. As a result, many companies are not willing to scale their infrastructure staff at the same rate that they scale the infrastructure itself. As a result, the IT staff is forced to take on more and more responsibility while working with dwindling resources. This is the very definition of that loathed phrase, "doing more with less."

This situation is one of the very reasons why people consider hyperconverged infrastructure. By massively reducing the variety of hardware and software that needs to be managed and maintained, the same number of IT pros can manage an increasing number of appliances providing all of the necessary services.

Hyperconverged Infrastructure and Your Job

Job security is a huge concern for many IT pros, particularly when waves of outsourcing have taken their toll. At first glance to many, the words *simplicity* and *efficiency* may appear threatening, but there are different ways to think about this.

Those most impacted by hyperconverged infrastructure are likely to be storage experts, because hyperconvergence pretty handily disrupts this area. Remember that many companies believe that storage is *the* most expensive and complex construct in the data center. Further, many storage pros end up spending a whole lot of time on relatively mundane tasks. If you're a storage expert, your company is probably hoping not to hire more of you anytime soon.

The goal here is to ensure that the services are still provided. That may not mean building out a SAN environment anymore. It may mean that the storage team should take the lead in helping to choose the right hyperconverged infrastructure solution — one that has the storage characteristics the organization needs. From there, if you are a storage pro, it will mean expanding beyond the storage realm into the world of servers and the hypervisor.

Solution Maturity

Hyperconverged infrastructure is the new animal in the data center jungle, but it's tearing through it in a big way. The overall packaging of hyperconverged infrastructure is relatively new. That is, packing everything into appliances and stacking them together is new, but the internal components aren't, depending on the solution.

There's a simple way to ensure that your intended solution can meet your needs: test the heck out of it.

Every hyperconverged infrastructure vendor on the planet will allow you to do proof-of-concept testing of their solution.

Your job is very clear. Test it. Put it through its paces and decide for yourself if the solution will meet your needs. Use *real-world* tests, not synthetic benchmarking tools to perform testing. After all, you won't be running synthetic tests in production; you'll be running real applications.

Eliminating the Modern Refresh Cycle

The data center refresh cycle is out of control and is way too tied to capital budget dollars. Many organizations tackle individual resources separately. For example, they replace storage in Year 1 and then every four years thereafter. They replace a few servers each year. They replace supporting equipment, such as WAN accelerators and SSD caches, every three years. This mish-mash of replacement cycles creates confusion and doesn't always result in an equal distribution of budget dollars.

What if you could implement a real rolling refresh process and simply add new appliances as new business needs dictate? And, as a reasonable replacement cycle emerges, you simply remove an appliance a year and cycle in a new one. **Figure 12-2** gives you a look at once such scenario.

The previous figure also demonstrates how you scale this environment. As you need to add new appliances, you just add new appliances. In the example shown in **Figure 12-2**, you start with three appliances in Year 1 and scale to four appliances in Year 2 and Year 3. In Year 4, you maintain four appliances but begin to cycle new appliances in. The entire process consists of racking and stacking new appliances as well as removing old ones.

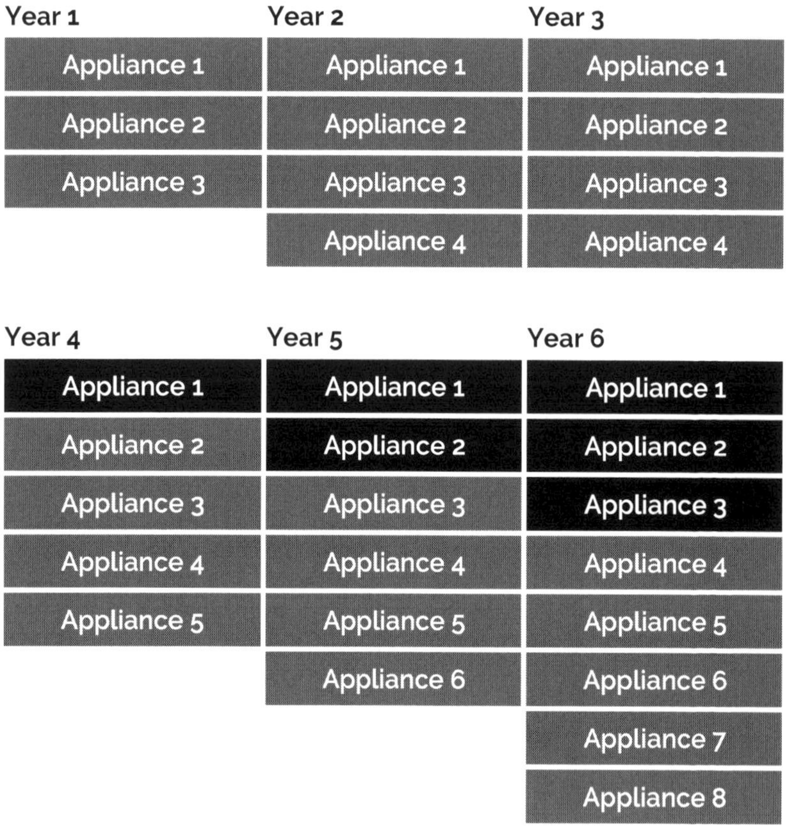

Year 1	Year 2	Year 3
Appliance 1	Appliance 1	Appliance 1
Appliance 2	Appliance 2	Appliance 2
Appliance 3	Appliance 3	Appliance 3
	Appliance 4	Appliance 4

Year 4	Year 5	Year 6
Appliance 1	Appliance 1	Appliance 1
Appliance 2	Appliance 2	Appliance 2
Appliance 3	Appliance 3	Appliance 3
Appliance 4	Appliance 4	Appliance 4
Appliance 5	Appliance 5	Appliance 5
	Appliance 6	Appliance 6
		Appliance 7
		Appliance 8

Figure 12-2: Hyperconvergence Refresh Process

This is a pretty new way to handle *every* infrastructure element. You no longer have to juggle storage, servers, and other supporting systems, including WAN accelerators, backup storage, and the other myriad appliances that litter the data center.

Obviously, though, you can't just tear everything out of the data center today and replace it all with hyperconverged infrastructure. Like most things, you need to be able to phase in hyperconvergence as it is possible from a budgetary perspective.

There are a couple of ways you can do this:

- **Introductory project** — If you have a new need, such as VDI or a ROBO modernization initiative, you can introduce hyperconvergence by implementing your new project on hyperconverged infrastructure. From there, on your natural replacement cycles, begin to migrate your other workloads to the new environment. Eventually, you will have fully phased out your traditional storage and server environment.

- **Make coexistence a priority feature** — Some hyperconverged infrastructure solutions enable peaceful coexistence with legacy infrastructure. For example, you're able to use existing vSphere hosts with the storage in the hyperconverged cluster. Coexistence with legacy infrastructure is not possible with all hyperconverged solutions, so choose carefully. Some solutions require you to implement hyperconvergence as a standalone silo of infrastructure.

Up Next

Understanding how to overcome inertia in order to make the right changes in your data center is really important. Also important are the economics behind the intended solution. That's the topic of our next and final chapter.

13

Hyperconvergence Economics: How It Impacts the IT Budget

Hyperconverged infrastructure has the potential to transform more than just the data center. By unlocking staff time and other resources, hyperconvergence can help your organization transform IT from a "keeping the lights on" cost center to a top-line revenue driver.

Focus on the Business, Not the Tech

We talked in Chapter 11 about the need for alignment between IT and the business. When the technology becomes too complex for IT to fully manage, or constantly requires the addition of new IT staff and skills, the focus shifts from the business to the infrastructure. In other words, people focus on building the infrastructure itself instead of enhacing what the infrastructure can *do* for the business. That can lead to a long-term problem and is one of the primary reasons that you'll start seeing misalignment.

The goal for most organizations must be to reduce the amount of "technical overhead." Just like other areas of the business, you reduce overhead to lower costs. You can do the same thing in IT with its own version of overhead. Reduction activities include simplifying administration, improving utilization of existing assets, and limiting the staff time spent on the care of feeding of the existing infrastructure.

The result is often reduced budgetary needs, and it can also mean that the company is better able to seize new business opportunities as they arise. That alone can have a dramatic positive impact on an organization's finances and the perception of IT's value to the business.

OpEx vs. CapEx

There are two kinds of expenses that you need to keep in mind when you consider data center economics:

Operational expenditures (OpEx) — OpEx generally aligns with the total cost of ownership (TCO) of a solution minus the initial costs. These are the ongoing expenses incurred when administering and maintaining what you've already purchased.

Capital expenditures (CapEx) — These align pretty closely with the initial cost of a solution and are often considered the one-time costs associated with a purchase.

Where Do the Savings Emerge?

This next question revolves around the actual source of savings. As you evaluate hyperconverged infrastructure solutions for yourself, these are some of the areas in which you'll need to focus to properly calculate cost and expense differential between your traditional environment and the hyperconverged environment.

- **No need to separate workloads** — By having the ability to consolidate workload silos into fewer or even just one silo, you can more easily increase overall utilization of what you're running. This harkens back to the early days of server virtualization, which carried increased utilization as a key driver for adoption.

- **Coexistence with existing systems** — Although this is not possible for *all* hyperconverged infrastructure solutions, the savings can be significant for those that can, in fact, coexist with existing infrastructure. In this scenario, you can more easily continue to use some existing systems (generally existing vSphere hosts) in conjunction with the new hyperconverged environment.

- **Reduction in electrical and cooling** — If you eliminate a bunch of servers, a slew of hard drives, and the accompanying infrastructure (such as WAN accelerators and SSD caches) from your data center, you will massively reduce your electrical and cooling costs. Less equipment translates to less power. Fewer moving parts means less generated heat, which leads to lower cooling costs. Don't discount these potential savings. They can be significant and are direct reductions on the operational side of the budget.

- **Reduced maintenance contracts** — You probably have annual maintenance contracts on everything in your data center. What would be the impact if you had fewer such contracts because you no longer required those hardware or software components in your data center? The likely outcome is that you would be saving money in the operational budget. As has been stated throughout this book, hyperconverged infrastructure has the potential to transform IT operations, including the operational budget.

- **Redirected staff efforts** — Less "tech" to manage means that the same or even fewer staff can manage it. There are fewer discrete skill sets needed in a hyperconverged infrastructure, meaning you can begin to redirect efforts toward goals that are more directly business-facing.

- **Training** – Top-down management (workload) vs. bottom-up (component) means less training will be required. Consider an iPhone. The iPhone handily replaces several distinct devices. But with the iPhone, there's only one interface to learn vs. the component-level management of a GPS, music player, camera, video recorder, timepiece, etc.

- **Operational and capital expense improvements** — You have seen a number of ways that the operational budget can be improved, but hyperconverged infrastructure can also have a dramatic impact on capital budgets. A lower cost of acquisition is just the beginning. As was mentioned previously in this book, you can implement a "rolling upgrade" paradigm. This can even out many of the spikes and valleys inherent in many capital budgets and enable easier scaling which doesn't rely on predetermined financial schedules.

Initial Investment Analysis

The initial investment in a solution is often one of the only financial elements considered by many organizations. That's because it's really important. They need to know how much a solution is going to cost right now. However, there are a ton of flaws inherent in this myopic model. First, you don't get a good feel for what the TCO will be for the solution. Furthermore, many traditional procurement models aren't even looking at what the business needs *today*. Instead, because of the way that most organizations have established budgets, they're buying the solutions that they need three or even five years from today. In other words, you're buying resources you'll have to grow into, not what you actually need today.

Let's examine this in more detail. In the **Figure 13-1**, you'll see two lines. The flat line is the purchased resource capacity at the inception of the current replacement cycle. The sloping line is the actual resource need in the data center. The shaded area is a zero return on investment zone.

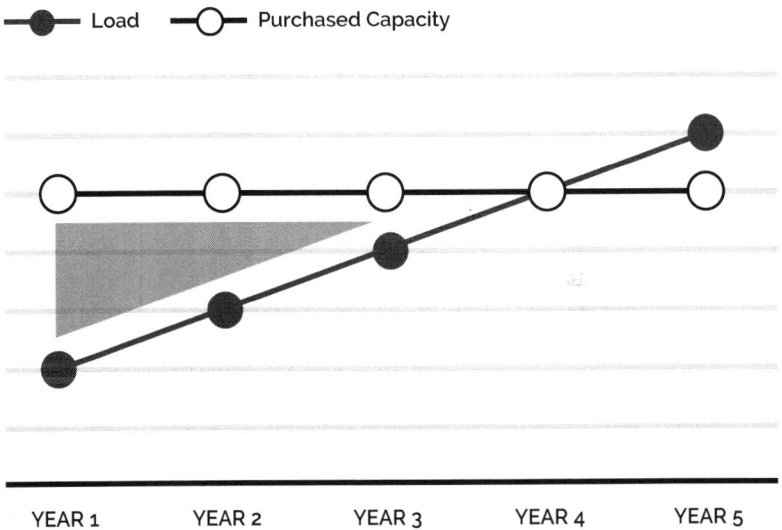

Figure 13-1: Load vs. Purchased Capacity in a Traditional Data Center Infrastructure

Economic Drivers for Technology Acquisition

There are three primary metrics that decision makers seek to satisfy when embarking on technology investments:

- **Return on investment (ROI)** — Any time you make an investment, you expect to see some kind of result from that investment. For example, if you buy stocks, you expect to see the stock price rise over time and eventually pay you back more than your initial investment. In the data center, an ROI can come in different forms. For example, you may be able to directly save money in another area by making a technology investment. Or, you may be able to save time on a process. In a business, saving time equates to saving money. The goal with ROI is to get back more than you put in. At the very least, your ROI should result in not moving backward.

- **Initial cost, or cost of acquisition** — If you can't afford the initial cash outlay for something, even the highest ROI and lowest total cost of ownership (TCO) won't matter. That's one of the reasons why the initial cost remains a top priority for many organizations.

- **Total cost of ownership (TCO)** — Everything you buy in the data center requires additional funding to maintain. This is beyond the initial cost of the solution. For example, for hardware and software alike, you generally have to pay annual maintenance fees. You may need to hire staff to maintain the solution. Additional staff costs money. You may incur installation charges or other fees. Everything that it takes to support a solution, including its initial cost, is considered the total cost that it takes to keep it, or the total cost of ownership.

For the first few years of this solution, there is massive waste in resource and budget. IT has to buy this way today because many data center solutions don't easily lend themselves to incremental scale. Instead, IT staff have to attempt to project the data center's needs for the next three to five years, and they won't always get it right. This is not a knock on IT pros; after all, stock market gurus often fail to predict markets, too. It's just the way things are based on the tools we have today.

Hyperconverged infrastructure solutions can help you break out of this cycle and more closely match data center resources with current business needs. By enabling granular and *simple* scaling, you can buy what you need today and, as the business grows, just add more nodes. Besides being technically simple, this also enables you to rethink the budgeting process. By not having to buy for needs that *might* be in place three to five years from now, you can focus on improving what the business needs *today*.

Your Financial Evaluation Criteria

As you consider implementation of hyperconverged infrastructure, **Figure 13-2** shows a simple worksheet you can use to help determine which solution makes the most financial sense — a traditional approach or a hyperconverged infrastructure.

1T = One-Time OG = Ongoing	Traditional		Hyperconverged		Difference	
	1T	OG	1T	OG	1T	OG
Server						
Hypervisor						
Storage/SAN						
Backup/Recovery Tools						
Disaster Recovery Tools						
WAN Accelerator						
SSD Cache						
Deduplication Appliance						
Power & Cooling						
Dedicated Staffing						
Other						
Totals			Savings			

Figure 13-2: Traditional vs. Hyperconverged Infrastructure Worksheet

That's a Wrap!

You've now been introduced to hyperconverged infrastructure, its use cases, plus the organizational and financial considerations for the technology. You've made it through the jungle of hyperconverged infrastructure and can move on to more harmonious locales. We hope your journey here has helped you better understand this important and growing segment of the technology market.

You can continue to learn much more about hyperconverged infrastructure by regularly visiting **www.hyperconverged.org** and **www.gorilla.guide**.